Digital Television at Home: Satellite, Cable and Over-the-Air

Using, controlling and understanding digital
TV technologies.

Gregory Dudek

Y-one-D books
http://www.Y1D.com

Published by Y-one-D books.
http://www.Y1D.com

Montreal, Canada.
Copyright © Gregory Dudek 2008
All rights reserved.

ISBN 978-0-9809915-0-5

Thanks to Krys Dudek and Stephanie Dudek for their assistance and unstinting support.

Library and Archives Canada Cataloguing in Publication

```
NAME(S)  : *Dudek, Gregory
TITLE(S):  Digital television at home : satellite,
         cable and over-the-air : using,
         controlling and understanding digital
         technologies / Gregory Dudek
NUMBERS:   Canadiana:  20089032713
CLASSIFICATION:  TK6678 D84 2008 621.388/07 22
SUBJECTS:  Digital television
```

1. Technology & Engineering 2. Television & Video 3. Satellite TV. 4. HDTV 5. Home theatre. 6. Video processing.

(08130808312)

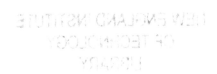

Table of Contents

15. INSTALLING THE DISH OR ANTENNA.................... 231

16. WIRING AND CONNECTIVITY.................................... 245

APPENDIX A.   BUYER'S PHRASE BOOK...................... 256

I. Introduction

What is digital television?

In this book we will look at how television works and how it is transmitted. In particular we will consider digital television (DTV), which refers to a new way of transmitting and displaying video images. Digital television is closely related to high-definition television (HDTV), which is a high quality system for displaying television images. Digital TV transmission and HDTV displays are not quite the same though: it should be noted that digital television *transmission* could can be, and is, used with older non-HDTV programming, as well as being used with HDTV. Digital TV is the basis of systems for satellite TV, cable TV and over-the-air TV that can work as well with standard definition analog televisions as it does with HDTV.

Television was once a pretty simple deal: you bought a TV set, attached a small antenna that looked a bit like pair of rabbit ears, and off you went. At least within a single region of the world, there was just one kind of TV and except for the really ambitious, there was no fancy setup required. The very absence of alternatives made it a unifying force in society, and a trivial choice for a consumer.

As the 21st century dawns, we are witnessing a major change in how telecommunications (TV, radio, telephone and internet) is delivered to individuals. There is an ongoing shift away from the previous standard of land-based TV and radio transmission. Such terrestrial broadcast is being replaced by two competing technologies: satellite-based transmission, and technologies based on wiring put in place for analog cable TV. Terrestrial digital television transmitted over the airwaves is part of the new landscape as well, but it has a lot of serious competition and it's not clear how well it will be able to fare. Cable TV uses co-axial cable (strands of wire wrapped around each other to share a common axis), and thus requires a physical cable to each place in which it will be used. Satellite-based transmission, on the other hand, works by beaming signals down from earth-orbiting satellites in outer space, and thus requires an antenna connected to each receiver. Each of these new technologies has advantages. In addition, there are also schemes for sending data via telephone lines (such as DSL) and local wireless (broadband) communications, but both of these currently offer lower maximum data rates than satellite or cable transmission.

Figure 1: A 1939 vintage RCA TT-5 television with a 5-inch screen. This set required a separate independent radio to receive audio. This photograph appears courtesy of the Early Television Museum, Hilliard, Ohio (earlytelevision.org).

While traditional television transmitted over land was developed roughly 100 years ago, and cable television was deployed in the 1960's, satellite television beamed directly for home use only became generally available in the 1990's. Despite its relative novelty, it has now become a major, perhaps the dominant, form of broadcasting. Back in the 1970's when Arthur C. Clarke (the writer who first proposed satellite communications) sequestered himself in Madagascar using satellite technology for personal communications, it was an exotic, eccentric and an impressive stroke of action possible only at exceptional expense. Today, college frat houses can pay for multiple satellite dish installations with money from T-shirt sales.

Figure 2: A family of C-band satellite dishes.

Traditional television as it existed for most of the 20th century was based on analog technology. Even computer output to video screens (using VGA for example) was an analog technology. This means that continuously varying voltage levels are used to convey information. Digital signal encoding is based on fundamentally different principles: signal values are quantized (i.e. labeled as having a specific value such as "voltage 122") and then these values are encoded as bit patterns (binary 1111010) and then only one of a limited set of alternative values, usually expressed using zeros and ones, are actually transmitted. Digital transmission has two fundamental features: error correction is very easy to implement robustly, and signals can be reliably encrypted. A third feature that arises indirectly from the fact digital technologies are much more modern is that the signals occupy less bandwidth, meaning more of them can be transmitted using the same physical conditions. How all this relates to the home viewer, and how it can be used, is the subject of the rest of this book.

In this book we will discuss what you need to get involved with this technology and how it works, in at least enough detail to allow you to make informed choices and to select and upgrade the technologies you use. We will also outline how to build or modify the software architecture for a digital TV system.

We will consider several kinds of systems and approaches for television data, but we will pay special attention to the **DVB** standard, an international standard for receiving satellite broadcasts at home, as well as **ATSC**, the North American variation of the approach (ATSC stands for the Advanced Television Systems Committee). Japan and Brazil have their own digital television standard called **Integrated Services Digital Broadcasting** (ISDB), which will not be addressed explicitly in the book, but the digital data encoding for ISDB is based on the same MPEG-2 underlying standard used for DVB, and is almost identical to it. These modern technologies allow robust terrestrial transmission, flexible cable TV, and permit the use of only a small dish antenna for satellite reception.

Cable versus satellite

Many television consumers face a choice between cable TV and satellite-based television, and the choice in not an easy one. There is no trivial answer to this question and we will start to examine it by considering the way the two technologies work.

Because cable (like DSL) only sends signals where they need to go, and it travels along a contained wire, it is much more energy efficient that other technologies where the signal radiates through the air. This leads it to possible cost reductions in terms of both the actual electricity used, as well as the physical systems that do the transmission. This is especially important for 2-way communications where the receiver sends data back to the transmitter (which happens when you browse the internet, for instance, or perhaps when a pay-per-view broadcast is selected). In addition, since the signals don't have to go all the way into earth orbit and back the way they do with satellite broadcasts, the time taken for a signal to go back and forth can be much shorter. The latency of sending a pulse of data back and forth over a satellite link is inevitably quite high even though the signal travels at the speed of light (about one quarter second each way from transmitting station to home, or vice versa). This is a long time for interactive applications.

Satellite television, on the other hand, requires less physical infrastructure on the ground. You don't need a huge network of wires spanning the countryside to provide satellite TV. You do, however, need launch capabilities and ground stations. The satellites themselves are very expensive to deploy, although they have gotten larger and cheaper as the technology has developed. The fairly recent **Anik F2** had a weight of 5900kg (13,000 lbs), as opposed to **INTELSAT I**, the first geosynchronous satellite (launched in 1965), whose weight was only 68kg (150 lbs). In the end though, the competition between satellite, cable and terrestrial broadcasting may depend on geography and consumer ease-of-use and convenience rather than hardware costs or power needs. It seems like in many cases, ease of setup per household trumps energy efficiency, latency, and even cost, since satellite broadcasting is the preferred solution for a huge fraction of the population.

When satellite technology was initially developed, both the transmitter and receiver required a very large and expensive dish antenna to operate. As the technology has matured over the last half century, every aspect of system has improved, but the most striking change has occurred in the receiving dish antenna. Instead of being a huge disk, satellite dishes for receiving transmissions can be as small as 12 inches (18 cm) in diameter for television: the size of a medium pizza (although 18-inch dishes (45 cm) are typically the smallest in commonplace use). These can be even smaller for radio-only broadcasting, and of miniscule size of the reception of satellite-based Global Positioning (GPS) signals. Satellite **direct-to-home** (**DTH**) television has gone from an exotic fantasy technology in the 1970's to a ubiquitous major force today. Its mid-term future probably depends more on social issues regarding the role of the Internet than it does on any remaining technical limitations.

Global positioning systems based on satellite technology have become a critical technology not only to navigate airplanes, but also for ordinary automobile drivers and to guide people hiking in the wilderness. Initially the technology was developed for military applications, and its full accuracy was withheld from ordinary users. As time has passed, it has gradually become more and more popular for a wide range for civilian purposes. As this popularity has increased, the technology has become cheaper, smaller and generally more effective. This is consistent with the increased and pervasive consumer presence of satellite-based technologies.

Due to the complementary nature of terrestrial, wired and satellite-based communication technologies, and their respective advantages and disadvantages, cable transmission and satellite based transmission systems will continue to co-exist with one another and each will have a role to play. It may be that there is a shakeout developing between terrestrial television, cable TV and satellite transmission for television and Internet connectivity, but it will be slow to arrive.

Three different media are used for transmitting television to the home: terrestrial (over the air) broadcast, wired cable TV, and satellite broadcast. The way digital data is sent over these media has several variations, but there are two large dominant standards for the protocols used (i.e. the data packaging): **DVB** and **ATSC**. The ATSC standard was developed in North America by the "Grand Alliance" of companies sponsored by the US Federal Communications Commission (FCC), while the DVB standard has a wider international presence. Each of these will be discussed and explained at greater length.

For many consumers, the need for access to over-the-air broadcasting is a given. Although this is more complex with digital broadcasting than it was in the past, some level of access is almost assured for all TV viewers. Many consumers face a much more daunting choice in selecting between cable and satellite. In this cases, the two systems provide similar features, but with a large range of subtle differences. No simple answer is possible regarding which is best, especially since both technologies are now evolving at a rapid pace, and thus the only sure thing is that things will be different again in a few years.

What is clear is that satellite-based television and radio have become major communications systems and are poised to continue expanding. To make informed choices about how to use these technologies, and what variations of them to buy, and how tools for them can be developed, it is important to understand something about how they work at a detailed level. Providing that information is the objective of this book.

Readers may wish to note that errata and supplementary information can be found on line at **http://www.y1d.com/DTVbook/links**

2. How TV works

Picture coding

Digital video signals still need to carry the same basic information as analog signals: pictures and sound that vary with time. The successive video frames in digital video are digitized (measured at discrete places on a grid), and each element of the grid (i.e. each PIcture ELement, or **pixel**) is expressed as a digitally encoded number (the name "pic-el" would have fit better, but it was already taken, with a slightly different spelling, by the pickled cucumber). Therefore, the brightness of the smoothly varying outside world is measured at specific points, and these points are the dots of light called pixels. The differences between differing image and video formats are related to how the grid of pixels is defined, how the numbers are expressed, and how the data may be compressed, or corrected if errors occur.

The number of elements that make up the sampling grid over the surface of the image determines its **resolution**. The specific standard resolutions that are commonly in use will be discussed later, but the concept is simple: more pixels mean more detail. Note that these grid elements do not have to have the same spacing in the horizontal and vertical directions. For good old standard definition (SD) video, used to digitally encode analog signals, the pixels cover a rectangular region of the screen (this leads to a 640x480 pixel standard definition video image with a width-to-length radio of 4:3). For high-definition (HD) digital video, the individual pixels are square.

In this chapter, we will discuss how a series of images that make up a video stream can be created on the screen. Both analog and digital video are based on the same core principles, so we'll discuss analog video as a precursor to understanding digital video.

Analog TV in a nutshell

Analog television was invented in the early 1930s, before the notion of digital signal processing, let alone the integrated circuit, had even been developed. The recognition for inventing television technology is shared between several engineers and scientists including Philo Farnsworth, who is often credited for developing the first complete television. Due to the intense competition to develop television, and the complexity of developing a practical system with no mechanical parts, many other figures played important roles including Paul Nipkow, John Baird, David Sarnoff, and Vladimir Zworykin. The assignment of credit for whose role was the most important remains a subject of dispute.

Figure 3: 1940's vintage black-and-white television set.

Analog television takes a set of pixel brightness values from a camera and converts them to a continuously varying electrical signal. This continuous variability of the brightness values is what makes the system analog. In addition, the placement of the brightness values along the horizontal lines making up the image is also defined in a continuous (analog) way.

These brightness values are then transmitted in some way, almost always using analog signals too, from a transmitter to a home receiver, at which point the signal is reassembled into a picture onto the screen. As always, the process of breaking down the signal and transmitting it (whether in a stored video tape or over the airwaves) leads to some amount distortion, interference and some number of errors in the information being sent. The key things that distinguish different analog formats (as well as digital formats) from one another is the manner in which they allow imperfections to be minimized or repaired.

Analog TV formats

Television was developed in the early 20th century as an outgrowth of radio. Radio is based on sending a loudness value from one place to another that varies with time (a one dimensional signal). Of course, the medium used for sending it is electromagnetic radiation, using what we have come to call the radio frequency. Sending television signal is more elaborate since it is necessary to encode a 2-dimensional picture, but still using only a single radio-frequency signal that varies in time, and this signal is the brightness at a single point that moved over the screen.

As an analogy, think of a pen moving back and forth across the screen in a fixed pattern at a fixed speed, moved by some repeating mechanical apparatus. The picture can thus be described using a one-dimensional specification of the amount of "ink" as a function of time from the beginning of the pen's path. After exactly $1/15^{th}$ of a second from the start of the pen's motion we know the pen will be at a specific position (say the middle of the screen), and ink sent out at that moment will make the middle of the screen darker. This knowledge of where the pen will be at a specific instant depends on the fact that we know that it takes exactly the same route over the screen every time. An amazing aspect of this approach is that to send a picture, all that needs to be transmitted is how much ink should be applied at each moment; the position of the pen is taken care of automatically since the sender and receiver have agreed on its motion in advance.

Of course in an actual tube-based television (a CRT), the pen is replaced by a stream of electrons, and it is guided to the right point on the screen by magnets that steer the electron beam. The result is very much the same as a moving pen across the screen, however, and continuous variation in the strength of the electron beam alters the brightness of the screen itself at the point that the beam is striking.

Figure 4: Close-up of a television image being painted on the screen. Scan lines are visible on the left, while the inset on the right shows the dots of line that make up each line. The display technology is selected to try and blur the lines and dots together.

The pattern of motion of the beam repeats itself every 1/30th of a second, and so it paints one picture after another. That is the essence of all television. It works to simulate motion because the human perceptual system is not sensitive enough to detect the difference between a flashed sequence of pictures and a smooth continuous motion. Note that the scheme depends critically on the sender and receiver both knowing where the beam (the "pen") is going to be at any moment, so that the specified ink shows up in the right place. In conventional television the motion of the beam is always a back-and-forth motion along successive horizontal lines at fixed positions. For **interlaced** video, which is the classic standard, the beam draws two successive versions of the picture, the second slightly offset from the first to better fill up all the space on the screen. The two groups of scan lines that together make up a full set of analog video lines are called **fields**.

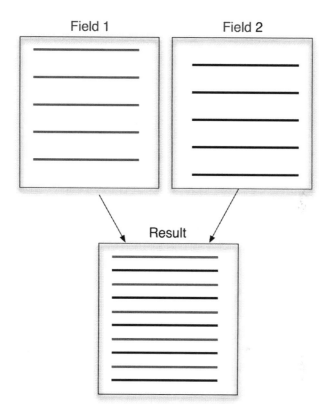

Figure 5: With interlaced video, two fields contributing are combined to make an image. One field covers only even numbered lines, and the other covers odd numbered lines on the display.

The different alternative standards for analog and digital television are determined by how many scan lines and pixels there should be, the refresh rate (how often the beam should repeat the pattern), and whether the scan is interlaced or progressive (i.e. non-interlaced). Of course, cathode ray tubes (CRTs) are being supplanted by liquid crystal displays (LCD) or plasma-based television sets, where the different locations on the screen are not painted by an electron beam, but instead are addressed digitally. When showing a standard definition television picture, however, these newer technologies essentially simulate the behavior of the good old electron beam.

A further issue relates to how *color* information in encoded in the signal. Colors in the real world depend on the amount of brightness at every possible wavelength of visible light, and this wavelength can vary continuously. Thus, there are an infinite number of colors possible. Fortunately for those who want to make televisions, the human eye has only limited sensitivity, and many different wavelength distributions in the real world are indistinguishable from one another to the human eye. An infinite number of different colors may be possible, but only a finite number of different colors are *distinguishable from one another*. Thus, a photograph of a banana appears to have the same color as a real banana, even though the specific wavelengths being reflected by the photographic inks and the banana skin are quite different. This phenomenon where the eye can't tell the difference between two wavelength distributions is called **metamerism**, and it allows TV to use a simple trick to reproduce colors. As it happens, the eye itself has only uses only about three different color measurements to guess the true color of point in the world. This is why a palette of three different screen colors, if chosen carefully, allow us to simulate our perception of almost any color.

Color television pictures are produced by using a screen made up of triples of small different-colored dots, very closely spaced beside one another, and these can be selectively turned on as needed (see Fig. 6). These dots are called phosphors, and they are turned on (briefly) when hit with the electron beam that paints the TV screen. With just three different dot colors, it turns out that almost any actual color can be simulated on the screen, at least well enough to fool the human perceptual system. In a similar way, a succession of snapshots displayed quickly one after another fools us into seeing smooth continuous motion.

When television was first invented in the United States, the idea of using differently colored phosphor dots on the screen had not been developed yet (and technologies for placing the dots and controlling the beam were not yet available). As a result, early television only used achromatic (black and white) brightness values with no color information and, as a result the television signals that were transmitted did not contain any color information.

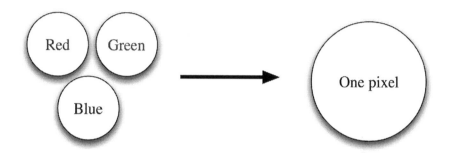

Figure 6: Color dots making up pixels

When the display technology for color TV was invented, it became appealing to consider sending out color pictures to home viewers. On the other hand, there was already a large installed based of home televisions that accepted the existing black-and-white TV signal. This was much like the advent of HDTV today: a more advanced technology needed a new kind of signal to be broadcast. Thus, at that time it was deemed necessary to find a way to transmit color signals without upsetting the large existing population of black-and-white television viewers. A backwards-compatible scheme had to be devised that would allow new television sets to receive a color signal, while still working with then-standard black-and-white-only sets. This involved encoding the signal needed for the color variations in "leftover space" (bandwidth/gaps) in the existing television signal, but there wasn't much of this space available. As a result, the color portion of the video signal could have only limited variability (or more precisely, limited bandwidth). In digital terms, there are fewer bits available to encode the color variations than there are for encoding the brightness changes. Luckily, the human perceptual system is not very sensitive to exactly where color lands when an image gets colored in. You can see this on antique-style Chinese vases where the blue dyes do not precisely match the boundaries of the line drawings, but people don't notice or mind.

Figure 7: Even though the shading colors on this porcelain plate do not match the outlines precisely, the effect is not unpleasant, or even noticeable, to the typical viewer

.If you snap a still frame of an analog television show, however, it is easy to note that the color placement looks a bit sloppy. This becomes especially noticeable and severe if an analog video signal is copied a few times, leading to increasing degradation. Duplicates of video tapes are almost unwatchable, which is why the broadcasting industry never worried about them as much as they do about digital duplication. Incidentally, in conjunction with the introduction of color transmission, the frame rate for North American television was dropped to 29.97 frames per second from a nice even 30 frames per second. This was done to facilitate the insertion of the color information because it led to a nice mathematical relationship between the existing audio carrier frequency, the video frequency, and the desired color data that had to be inserted.

There are several major standards for analog television, depending on where in the world you live. The **NTSC** standard (National Television System Committee) is used on North America, Western South America, the Caribbean, and Japan. The **PAL** standard (Phase Alternation by Line) is used in most of Europe (except France), as well as China, Brazil and various other countries from Afghanistan, to Zimbabwe. The **SECAM** (SÉquentiel Couleur À Mémoire) standard that was developed in France is used there and in former French colonies (especially in North Africa) as well as in Russia and much of Middle East (where PAL is also used). At 525 lines per image and 29.97 frames per second, NTSC has the highest frame rate and lowest resolution of these systems, while PAL and SECAM each provide 625 lines per image, but at a lower rate (25 frames per second). While there are weak technical excuses for using one standard over another, a large part of the explanation for their existence, especially in the case of SECAM, is to protect local television manufacturing and broadcasting infrastructures, or to provide a cross-cultural barrier. If the country next door uses a different video standard, their programming is easier to ignore, and their sales and production infrastructure is easier to ignore or displace locally.

Notably, SECAM and PAL work the same way as one another with respect to the luminance (non-color) information, and thus SECAM video signals can be viewed on PAL equipment in black-and-white. In addition, due to the relatively small SECAM market, no SECAM DVD technology was ever developed, so even in regions where SECAM is in use much of the technology, including all DVD equipment, also supports PAL.

The signals that bring programming can vary either in total strength (i.e. amplitude) or else in the difference between the strongest and weakest signal, as a result of the transmission process and how close you are to the transmitter. As a result, many video systems include circuits for automatic gain control (AGC), which means they automatically change the amount of amplification to enhance and adjust the signal to get a consistent variation between the whites and the blacks in an image. In simple terms, this corresponds to normalizing the signal so that the minimum value is the darkest available, and the strongest signal is the brightest. There are some tricky aspects to this to allow for the fact that some scenes might actually be mainly dark (like night-time shots), but that's the basic idea. It turns out that by taking advantage of quirks in how analog AGC circuits operate, some types of copy protection can be implemented, as we will discuss in a later chapter.

Subtitles, closed captioning and teletext

Analog television, as well as digital programming, can have descriptive "closed caption" text associated with it. This is most commonly used to provide subtitles for hearing-impaired viewers, or for multilingual translations, but is can also be used for other purposes such as providing program guide information.

With NTSC-formatted analog video (and related standards like PAL), there are more scan lines in the signal that are actually displayed on the television screen. One or more of these "spare" scan lines is used to store the closed caption text that can be decoded and displayed by suitable televisions (or software). Specifically, the text is stored on scan **line 21** of the analog video signal during the vertical blanking interval. North American *digital* broadcast television also stores closed caption in line 21 (as well as elsewhere) to provide compatibility with analog televisions. One advantage of storing the text within the video signal is that it is automatically captured by analog recording devices (such as VCR's), without them having to do any special processing even know it is there. On the other hand, if the video picture itself is digitized or transformed, the subtitles are lost since they are not actually part of the picture itself.

This "line 21 data" can provide various different text "channels" that are associated with a single program. Special characters that are inserted into the closed caption strings specify the data for these channels, and the position of the subtitle text on the screen. The first video field of a frame contains four channels, referred to as CC1, CC2, T1 and T2. The second of two fields making up an NTSC frame contains another four channels CC3, CC4, T3 and T4 as well as **Extended Data Services** (XDS) information. XDS information is used to send ratings information for the **V-Chip** that can be used to censor programming, to send time-of-day information to automatically set clocks, or to provide basic programming and station identification information. XDS support included with many recent VCR's allows them to automatically know what time it is, and thus avoid the legendary "blinking 12:00" seen in many households where the owners of older devices were unwilling or unable to set the clock themselves.

Teletext is an alternative system for sending text information along with video. It is used primarily with PAL and SECAM broadcasting formats, but is compatible in principle with NTSC as well; it just isn't used in North America. Teletext is also based on inserting digitally encoded text information into the vertical blanking interval and it is used for various kinds of information, such as stock quotations, in addition to subtitles. Depending on the broadcaster and region, it can use a subset of lines 6 through 22, and 318 through 335 of the video signal, and achieves about 40 bytes of data per line. This allows successive **pages** of data to be sent one after the other in a repeating loop, and it is the task of the television to store these and provide an interface to the user so that they can select the page they want to see.

Other related systems such as the "TV Guide On Screen" or the "Video Program System" (VPS) have also been developed to use data in the VBI signal to send schedule information, often to allow for automated recordings.

The situation with captions and subtitles with digital video is a bit more complicated, partly due to the needs for backwards compatibility. Analog closed caption is legally required in many countries including the United States. The legal landscape for digital captioning is much less uniform. Digital video can include caption information as part of the set of digital packets that make up the transmission, as is the case with most MPEG-based programming. In addition, digital transmission sometimes includes traditional analog-style caption data to allow for backwards compatibility with older devices.

Analog copy protection

Digital television makes copy protection a serious endeavor, but even before digital standards appeared, there was a desire to protect analog video from unauthorized recording. The trick that was developed to do this exploited the differences between recording devices and television sets, so that programs could be watched, but not recorded. Schemes to achieve this were developed, and are discussed in the chapter dealing with Conditional Access Control and DRM (page 152).

Color and brightness information

In most picture storage systems, even including photographic film, color image data is stored using three values. These can be regarded as three "paint cans", three light bulbs, or more formally as three "basis functions". In addition, the combinations of the brightness of the three colors being combined at each point add up to the total brightness of that point. We have already discussed how the human perceptual system has limited sensitivity to color variations and, in general, is more sensitive to changes in brightness than to changes in color. The result is that there are many different ways of expressing the combination of brightness and color at each point in the image and, in general, we can optimize performance by sending less bandwidth on color information than on brightness information, since the eye notices it less. Almost every video encoding scheme uses three different components to send color picture content, but these three data components can be used in various ways known as **color spaces**.

One scheme called **RGB** sends independent signals for the amount of red (R), green (G) and blue (B) light at each point. This is conceptually simple, corresponds to what the display screen is actually doing physically, and leads to simple algorithms and a simple circuit. As a result, RGB encoding is used in numerous computer applications. The problem with RGB is that it "wastes" bits since on a high-resolution image (especially a moving one) we could get away with sending less information about color variations, and a human who isn't equipped with a magnifying glass would never notice the difference. To do this, the average brightness or **luminance** is computed at each point, and the transmitted signal uses frequent samples of the brightness, but less frequent measurements of the color variations. Once the brightness is known, the color can be computed with just two more measurements (since the three color channels add up to the brightness). Several different approaches for computing the luminance and color have been invented, and the standard often used for digital TV is **YCbCr** (or as **YUV**). In that scheme, **Y** refers to luminance, **Cb** refers to the blue part of the color signal, and **Cr** refers to the red part of the color signal. The Cb and Cr signals can be calculated at each pixel by subtracting the **B** or **R** parts of the **RGB** signals from the **Y** luminance signal. This same color space can also be used for analog signals, and in that case it is often referred to as **YPbPr** (and in fact the names are often confused and used interchangeably, along with variations such as gamma-corrected **Y'CbCR** or **Y/Pb/Pr**, and sometimes pronounced in English as "yibber" or "yipper").

The color variations (without luminance) are called **chroma** and the technique based on measuring them less often than luminance is called **chroma subsampling**. The frequency of subsampling is calculated in proportion to how often the luminance is sent. The advantage of subsampling the chroma signal is that the video signal becomes more compact, and can be stored and transmitted more efficiently. As a baseline, we assume we have 4 samples of luminance, and then if each of the two chroma parts is sampled half as often as luminance we get a **4:2:2** signal. Other common sampling frequencies are **4:1:1** and **4:2:0** (where zero has a special interpretation, and does not mean omitting sampling of one component). The 4:1:1 sampling leads to noticeable color artifacts in some special situations, but despite that it is the standard used for DVCAM video recording, consumer DV, and DVCPRO formats. If all parts are sampled equally often, the sampling is **4:4:4**, which is about the same as just sending the RGB signal.

Comb filter

Many analog televisions include a **comb filter**, and this mysterious item is often cited an important feature in advertising. It sounds like some kind of grating that picture gets poured though, or perhaps a way to dealing with unkempt hair. In fact, comb filters are small circuits that improve picture quality by attempting to compensate for tradeoffs in the way analog television works when transmitted used component video cabling (or received over the air). They are useful even for digital television devices that need to display stored analog programs (such as from DVD or tape).

A comb filter is, in essence, a circuit that adds together different scan lines (or frames) of the picture to better separate the brightness signal from the color information. When video is transmitted over the air, or using a composite video cable, the brightness and color signals are mixed together and can interfere with one another. When there is no motion in the scene, adding together pairs of lines can help determine where the brightness changes should go and where the color variations are relative to them. This is a result of the technical details of the color encoding, the high frequencies that are used to encode it, and the fact that the color signal will change even when the image contents (and brightness) do not. (The name of the comb filter comes from the appearance of the mathematical function this addition corresponds to, when depicted with respect to various signal frequencies.)

The best kinds of comb filters are called "motion adaptive" (or equivalently "3D"). These fancy variations work by first figuring out how much motion there is in the scene, and adjusting to amount of filtering to do to get a good result.

Digital resolutions and formats

The term "digital television" can have two different meanings pertaining to either the picture or the transmission system. There are formats for the television signal that are analog (most commonly in the form of **standard definition**—SD), like NTSC or PAL, but which is transmitted over digital signal transmission systems using DVB and MPEG encoding. In such a case the analog nature of the signal is only reconstructed as the last step before it is fed into the TV display. Alternatively, the display itself may accept a digital signal, and such signals are encoded at a higher resolution (**HDTV**) than older analog signals. In the case of HDTV, the signals are always transmitted and stored digitally as well as displayed digitally.

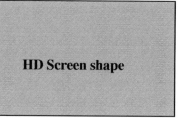

Figure 7: Standard definition and High definition screen aspect ratios,.4:3 and 16:9.

High-definition digital television is a term that is used to refer to a collection of inter-connected technologies: the transmission mechanism used to send the picture, the data encoding method used to digitally encode the signal data, and the scanning and sampling methods used by a television set to display a picture on the screen. Most of all, it refers to the combination of the number of pixels on the screen and the way they are encoded. While we will discuss several of the technical issues later, it is worth saying a few words now about HDTV displays operate. In short, they provide two advantages over analog television, aside from clearer signals: larger numbers of pixels and faster refresh.

The most noticeable thing about HD televisions, as compared to analog ones, is that the size of the screen is different. Analog television (standard definition TV) has a screen that is 4/3 times as wide as it is high, thus the aspect ratio is 4 to 3, or 4:3. The aspect ratio is the name for the ratio between the width and height, i.e. the shape of the screen. HD television, on the other hand, has a wider screen than SD with an aspect ratio of 16:9, more like a movie theatre. This is illustrated in the figure above.

The number of pixels on digital televisions can vary quite a bit, but it is typically at least 720 lines (rows) of pixels, with 1280 pixels along each horizontal line (which gives a total of almost a million pixels). The 720 pixels are what you get if you start at the left-hand end of the display and count dots of light as you move from top to bottom. In addition, digital television (like analog TV) can be shown using either an **interlaced** or

progressive display. In a progressive display successive lines are put on the screen one after the other. In an interlaced display lines are painted alternating between a field made up of only the odd-numbered lines, followed by second field of only the even numbered lines as introduced previously (page 12). For the record, progressive display is occasionally referred to as sequential scanning. Clearly if the fields (each holding half the data) in an interlaced display arrive at the same rate as the frames holding a full screen of data on a progressive display, it means the lines on an interlaced display get refreshed half as frequently. For this reason, when two displays with the same number of pixels are compared, a progressive display device is considered better quality than an interlaced display. The rate at which these fields or frames arrive is typically 60 times per second. These display types are commonly referred to by taking the number of lines on the display, adding the letter "**p**" or "**i**" for progressive or interlaced, and then optionally adding the frame rate. So a device with 720 lines, a progressive display and a frame rate of 60 frames per second is denoted as **720p60**. The details regarding the number of pixels used with different formats are presented in full detail in the appendix on Standard Video Formats. The 720p60 and **1080i** formats are two of the most commonly used, with 720p being the standard for over-the-air broadcast in North America.

While standard definition remains the dominant viewing choice the world over, HDTV is making steady inroads (see *The end of terrestrial analog* on page 48). When color TV was introduced, it was developed to be backwards compatible with existing black-and-white television, and it gradually completely forced out black and white from the market TV by virtue of being more appealing to consumers. The marketplace and the consumer led the transition. In contrast, the transition from analog to digital television is being catalyzed by legislative action that is forcing analog television off the air and replacing it with terrestrial HDTV

> Quick summary:
>
> Broadcast HDTV comes at **720p** resolution. This is about the same as **1080i**
>
> **1080p** is the best quality of HDTV, but is usually only available today from video game machines like the Xbox or PS3.

broadcasts in the United States and most other countries. There are several reasons and justifications for the introduction of this legislation, some of which are explicitly stated, and others that are simply matters of conjecture and speculation.

What is clear is that additional radio spectrum will become available with the transition for digital TV, since it uses the available spectrum more efficiently. It is also clear that a lot of money is going to be made, if only due to the fact that almost all consumers will be compelled to spend large amounts of money to upgrade in the long run.   In addition, while analog broadcasts are very difficult to protect from home recording and duplication, digital HDTV broadcasts have digital rights management (DRM) protocols built in. While these DRM protocols for HDTV can probably be cracked, it is inconvenient for consumers to do so, and generally requires commercial tools to make it easy. Further, the United States is taking a leadership position in pressuring all countries to follow it in adopting legislation that would make this illegal, and thus prevent the trade in such tools, even when they can be constructed at home (this topic is discussed further on the chapter entitled "Copy protection, Conditional Access and Piracy"). In short, there are many reasons for going digital, and many parties in addition to consumers with a vested interest in seeing it happen.

Satellite or cable broadcasting could remain largely analog, but will probably feel tremendous pressure to increase both the proportion and total amount of digital programming. Of course, the huge existing volume of older "legacy" programming, such as television shows from the last century, will always be limited to standard definition, even when transmitted digitally.

In the United States, terrestrially broadcast digital television occupies largely the same portion of the electromagnetic spectrum as traditional analog television. Due to its digital nature, if the packets making up a program are too badly degraded, no signal whatsoever will be viewable. As a result, in locations with poor reception digital television may not function whereas analog television might still deliver a degraded signal.   On the other hand, if a digital signal can be received, then it will almost always appear crystal clear without many of the artifacts and the video noise usually associated with analog television reception.

Digital compression formats

When digital video data in encoded in the form of packets, the encoding process can take many different forms. In general, it is possible to avoid storing the value of every individual pixel. The essence of data compression is the fact that the value of one pixel allows the next one to be guessed with some high probability; in such as case, all that needs to be stored is the adjustment to the predicted value. Video compression formats are almost always **lossy**, meaning they compress the data at the expense of some loss in quality. While the set of compression and encoding options is too large to be listed here, the main outlines are all that matter for most practical purposes.

Video

Video compression is used primarily to allow more data to be sent when the capacity to send information is limited. The receiver then needs to decompress the data to get a usable picture. There is a diverse set of different strategies for accomplishing this compression, but almost all compression methods for video adhere to a few basic principles. The most critical principle is to avoid transmitting data that can instead be figured out from what is already available. For example, if an image has almost uniform colors, then given the colors on the odd scan lines, the colors of the even scan lines can be estimated without sending the data, or perhaps only sending the an extra tidbit saying that the color is almost uniform (as opposed of made out of tiny stripes). Using such tricks, of course, requires the encoding system and decoding system to agree in advance on what needs to be sent, and what will simple be inferred. So video compression cannot be used without decompression, the combined set of program code to accomplish it is usually called a **COder/DECoder**, or more commonly a **CODEC**.

Note that the way the video data is compressed is distinct from the way the pixels are eventually stored or transmitted. Compression refers to the mathematical transformation from pixels to compressed pixels; how these are stored once the transformation is made is a separate issue. Thus, a given compression scheme can be used with different video **container formats**, just as a given English phrase can be written on the pages of a book, on index cards, or scribbled on a blackboard.

Video compression can be applied to single frames of video independent of one another, but it is generally much more efficient to encode one frame relative to the ones before or after it. We can think of data being sent from the coder to the decoder, and the decoder can often use preceding frames to guess most of the current frame. For example, if the scene being shown does not change at all, almost no data needs to be sent for successive frames except to indicate they are the same as the one that came before them. More generally, if only some parts of an image move, it may be possible to send only these changed parts of the image. This, however, requires a compression program smart enough to automatically detect what has changed from one video frame to another. The underlying principle behind this is called motion detection or **motion compensation**. Several different methods are available to do this, and the sophistication of the approach is part of what distinguishes older MPEG-2 video compression methods from the more effective and modern MPEG-4 standard. In fact, some of these compression systems can compress frames using information about not only preceding frames, but also future upcoming frames (and hence they need to peek ahead in the video sequence). MPEG encoding generally works with differences between several video frames at a time, referring to these as a GOP (for group of pictures).

Both MPEG-2 and MPEG-4, as well specific compression systems like **H.264** and **XVid**, allow for variations in the compression process even within a single video sequence. In fact, MPEG-2 and MPEG-4 are "standards families" that specify how video should be handled, but with enough generality to permit variations in how the video is compressed, while still adhering to the specification. MPEG-2 and MPEG-4 are internationally regulated standards that are used for video data, and specifically for DVB and ATSC broadcasting. The MPEG-4 specification has many parts related to different types of compression (audio, video, file formats) and some of these are further subdivided into "profiles" are describe different sub-classes of application (heavily compressed low bandwidth, high bandwidth). The **H.264** and **XVid** CODECs are popular MPEG-4 compliant algorithms used for video data. H.264 is the official name assigned to a high-compression CODEC by the International Telecommunication Union (ITU), while the same algorithm is also called designated **MPEG-4 part 10** by the MPEG organization. XVid is an open source implementation of MPEG-4 encoding and decoding (although its commercial use may be restricted by patents in some parts of the world).

A common question is: "which compression system is best?" In fact, there is no simple answer, any more than there is to the "which computer is best for all buyers," or "which religion is best?" Factors that come into play are the speed of the CODEC for encoding video (for those who want to make recordings), the speed of decompression, the size of the resulting compressed file (i.e. the compression ratio), as well as commercial restrictions and licensing requirements. Some of the most efficient CODECs in terms of compression are also very computationally intensive, and demand costly high-powered processors to be bearable. Some CODECs work very well for static content, but do badly when there are large amounts of motion in the scene.

Audio

Essentially all television broadcasting is a combination of a both video and audio. Like video, audio data can be compressed in many different ways. In addition, as with video, the compression method is distinct from the data storage container format, although the two are often confused.

Perhaps because the size of an audio track is far smaller than the corresponding video stream in a single program, the choice of audio CODEC for audio/video programming seems less confusing for most people than the video case. On the other hand, as with video data, it is difficult to make objective statements about which encoding scheme is best under all circumstances. In fact, perhaps because audio data is less tangible, there seems to be less agreement over what actually constitutes good quality for audio compression and encoding in any particular case. On the other hand, a good rule of thumb is that newer CODECS generally can generally provide smaller files (i.e. more compression) for a given perceptual quality, if the encoding settings are tuned properly.

MP2

The **MPEG-1 Audio layer 2** format, known informally as **MP2**, is a high performance audio coding system developed in the early 1990s. Note that this is not the same as the MPEG-2 standard used for video, and in fact MP2 audio is used within MPEG-2. Its official designation by the International Organization for Standards is **ISO/IEC 11172-3**. Is it based on two fundamental principles. This first of these is common to all compression schemes: eliminating redundant information (that is, information that can be guessed from what came before does not need to be transmitted). The second principle of MP2 encoding is the elimination of audio content that humans will not be sensitive to, even if it has otherwise genuine information. This **psychoacoustic model** is based on detailed analysis of human sound perception and is used in some form by most modern compression systems. For example, for a short time after a loud percussive sound (like a drum beat), we are not sensitive to weak sounds that follow it, and so they can be discarded without altering our perception of the recording.

MP3

The **Mp3** audio format is not used directly for broadcast television, but warrants mention as the most prevalent method for audio compression. Its official title is **MPEG Audio version 1 layer 3**. It is based on the MP2 CODEC and applies additional transformations to the audio data to achieve slightly better compression, arguably at the cost of slightly lower quality.

AC3

The AC3 format was developed by Dolby Labs, and is also known as **Dolby Digital**. It is a high-performance audio encoding system that allows for multi-channel audio (meaning multiple speakers for the same audio track). In general, specifically encoding multiple channel audio in one combined package of channels is much more efficient than encoding the channels (different speakers) as different independent audio programs, since the channels share information. A two-channel sound encoding is needed for stereo. Surround sound, or 5.1-channel audio refers to the use of five regular speakers: front center, and two front and rear sides, and well as a special low frequency-only speaker acting as a subwoofer. Typically AC3 audio can carry one to six channels, and uses a data rate of 96 kilobits/second for single channel monophonic sound, 192 kbits/sec for stereo, and 384 or 448 kbits/sec for 5.1 audio. An extended version of the format called Dolby Digital Plus allows for additional audio channels (to support 7.1 channel audio), has higher bit rates, and has better encoding performance for low bit rates.

In addition to broadcast applications, this format is used for DVD video recordings, and Blu-ray disks.

DTS

The DTS audio format, sanctioned for use in both DVB audio as well as Blu-ray disks, is aimed particularly at surround sound audio reproduction (i.e. 5.1 audio). It has similar features to AC3 audio, also using perceptual coding to achieve compression, and also having many variable parameters to trade-off compression for quality. By and large, it appears to be little-used for television broadcasting at present, but it is used on Blu-Ray Disks.

AAC

The MPEG-4 container format allows for the use of several audio CODECs including the **Advanced Audio Coding** (AAC) standard and **High Efficiency Advanced Audio Coding** (HEAAC). AAC is used with DVB video in three variations: regular AAC, MPEG-4 HE AAC and MPEG-4 HE AAC version 2 (v2). AAC, HEAAC, and HEAAC v2 provide increasingly better compression for roughly the same perceptual sound quality.

All versions of AAC generally claim to have better performance with respect to the tradeoff between sound quality and compression than older formats that were defined for use with MPEG-2, although AAC is more demanding with respect to computation and has a substantially more complex algorithm. AAC itself allows for substantial variation in how it is applied, since its performance is controlled by many variables. This allows its performance to be tuned to different types of audio application, for example classic music (which is acoustically demanding) as opposed to human speech (which tends to make very limited demands on acoustic performance).

Audio Transmission

The DVB protocol using MPEG-2 allows audio to be transmitted using two different encoding methods, specifically MPEG Audio version 1 layer 2 (i.e. MP2), DTS audio and AC3. Since AC3 is a more recent addition, some broadcasters may also choose to broadcast programming accompanied by both MP2 and AC3 audio, in order to support older receivers. DVB video using MPEG-4 also permits the use of AAC audio, although this is not in wide spread use yet.

The ATSC protocol is more consistent regarding audio coding. It requires the use of AC3 exclusively.

3 Telecommunication Principles

Digital television is based on creating a digital data stream, that is, a sequence of ones and zeros that encode all the material we are interested in. In this chapter we will look at how to use radio frequency communications to send a signal, regardless of its particular content, from one place to another. This telecommunication process uses the same basic principles that Marconi exploited at the start of the 20th century to send the first radio messages.

To fully understand modern broadcast communications, and all the various frequencies that need to be adjusted, we need to get just a little bit of telecommunications theory under our belts. We don't need a lot of technical details, just enough to grasp the basic concepts. This proves to be important for some practical applications since the various frequencies used for some parts of the system need to be manually specified for some kinds of equipment.

How signal transmission works

Satellite broadcast television, like terrestrial television and regular analog radio is based on transmitting electromagnetic signals. In the cases of television, the transmitted signals are used to vary the brightness of the dots on the television screen that make up the picture. They are also used to vary the form of a loudspeaker's vibration to produce sound. Thus, the crux of the transmission process is the ability to send various frequencies from the transmitter to the receiver and have these describe the signal strength (i.e. amplitude) of the signal.

The frequency of a wave is the rate at which is oscillates. If you watch a wave passing, it is also the rate at which successive waves pass, for example the rate at which successive crests (humps) of a wave pass by (see Fig. 8). A sequence of waves is sometime called a wave train. The wavelength of a wave train is the *distance* between one crest and the next, but we can equally well measure the distance between any fixed point on the wave such as the trough at the bottom, or the mid-point.

The *height* of the waves is the **amplitude**. This is a more general term than "height" since it also applies to other kinds of waves where the notion of height is not intuitive, such as light (where amplitude refers to brightness), sound (where amplitude is loudness), or electricity (where amplitude can be either voltage or current).

Recall that the rate of variation of a signal is referred to as its frequency: how often it stops and start. We can even refer to the frequency of the beating of your heart, about 72 beats per minute which equals 1.2 beats per second, or 1.2 **Hertz**. Since electromagnetic radiation (radio, microwaves and light) is a form of wave, we can also consider the corresponding wavelength for a signal at any frequency. In general, the Greek letter λ (Lambda) is used to refer to wavelength. Although it may be less familiar, the general rule is that the wavelength λ of a signal moving in any medium is given by the speed of its motion c divided by its frequency f:

$$\lambda = c / f.$$

The **period** T is a signal is the time taken for one wave to pass, equal to:

$$T = 1 / f.$$

Here is a real-world example. For a tsunami wave moving in the water we can estimate the distance between waves (the wavelength) if we know the frequency of the waves and the speed of the waves' motion in water. For a typical tsunami in shallow water, the frequency is 1/60 Hz and the speed is

166 m/s, thus the wavelength is almost 10 km. Using the same formula, we can also compute the speed (166 m/s or 544 ft/s) from our knowledge of the wavelength and frequency. Note the frightening fact that this speed is a lot faster than you can run; in practice, actual tsunamis vary in speed depending on the depth of the water, and can have periods of 2 to 30 minutes, wavelengths from 8.5 to 200 km (5 to 124 miles), and speeds in deep ocean of up to 1000 km/hr (621 mph). For electromagnetic radiation (including radio waves and microwaves), the speed we need to use in this formula is the speed of light, 300 million m/s (671 million miles/hour).

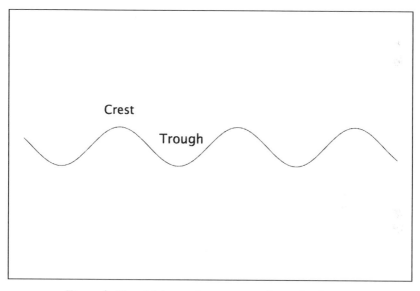

Figure 8: Sinuoidal waveform, crest and trough of a wave.

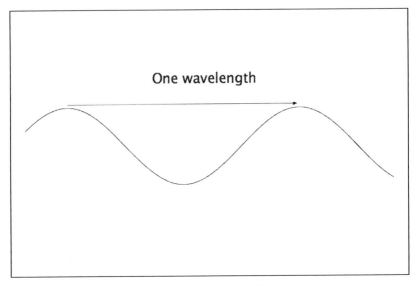

Figure 9: Wavelength

Carrier modulation

If the transmitter and receiver were connect by a simple piece of wire (as they are within the some entertainment system), then these signals of interest can be sent trivially as variations in voltage or current. The signal we are interested in is precisely the same signal that we sent, presto! In general though, we can't do that due to physical restrictions on the sending process. There are several reasons for these restrictions. The essential reason it that the signals we want to transmit (such as music frequencies) are almost never made from the same frequencies as the ones that we can efficiently send (such as microwave radar frequencies). The factors involved are related to the fact that any single signal can be viewed as having different frequencies within it (for example high notes and low notes). A technical version of this involves the Fourier transform of the signal being sent, but all that really matters in practice is that any normal signal can be seen as being made up from many different frequencies that are combined together. When signals are broadcast, we need to use antennas and amplifiers to send and receive the signals. Antennas, in particular, operate best at very specific frequencies (related to the wavelength of the signal being transmitted and the physical size of the antenna). As a result, although we want to send a wide range of different frequencies corresponding to whatever our data is, the particular set of frequencies that can be efficiently transmitted or received using a specific antenna is more limited. In addition, we may want to send many different signals at once, and we can't reserve all the different required frequencies for each one. The solution to this puzzle is a trick known as **modulation**, and it is the basis of almost all radio communication.

Consider the simplified case in which our antenna (and associated gear) is good for transmitting and receiving precisely one specific frequency, and that's the only frequency we can transmit at. We'll call this the "carrier signal" since it well have to, somehow, carry our data. Let's also assume we have a "data signal" we want to send which has a lower (perhaps variable) frequency.   This "data signal" represents the actual message, be it either text or music, that we wish to send.   It fact, to construct a really, really simple example, we can consider an imaginary digital signal we would like to transmit where the frequency itself doesn't matter at all, we just want to get a few individual *digital* bits transmitted. For example we want to send some specific bit pattern that might change from time to time (sometimes 0101, later on, 0110, etc.). The problem is that our antenna in this example can only send the one fixed frequency.

In this very simple case, we can turn the carrier signal we are transmitting completely on or off after each full cycle to indicate successive bits. Then if there is any carrier, that means the data bit being sent is a one, if there is no carrier present at all, it's a zero. This is roughly how Morse code works. The bit pattern 1101 would be sent by starting with some signal (i.e. a one), then some more signal (another one), then no signal at all (a zero), and then some signal (a one). You can imagine transmitting a signal in this primitive way by flashing your bedroom lights on and off to signal a neighbor ("SOS, MY TV IS BROKEN").   Note that in the case of the bedroom lights, just like our antenna, the light bulb only transmits a fixed carrier frequency of light (i.e. the lamp color), and the signal of interest is hidden in the "on's and off's" of that fixed carrier signal.

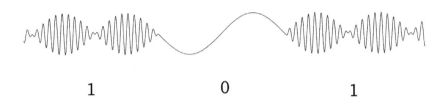

1   0   1

Figure 10: Simplified example of a low frequency carrier modulated using a second frequency whose presence signals the simple digital bit pattern 101.

More generally, modulation can be accomplished by changing the amplitude (the strength or loudness) of a carrier signal according to a timing pattern than encodes a data signal. In the most extremely simplified case like the example above, the data signal might correspond to just "on" and "off" values. More generally, the data we are sending can encode specific numbers via specific amplitude values, for example the carrier strengths "off",

> **Amplitude Modulation** is based on sending data by altering the transmitted signal strength.

"weak", "medium", and "strong" could be used to send any digit from one to four, instead of just ones and zeros. The set of possible values that we read off the signal can be seen as the *alphabet* for our transmission. In this way, the signal being transmitted can be described as being made of two different frequencies: the carrier frequency, and the (somewhat variable) frequency that makes up the actual data of interest. This kind of technique is called **amplitude modulation (AM)**, and it is the basis of AM radio. The music is used to modulate the carrier, and the carrier signal is the station frequency you dial in to on the radio. The illustration below shows how a bit pattern, in this case 010, can be encoded by changing the amplitude over different discrete units of time. In this sample case, the binary digit "1" is indicated by reducing the amplitude of the signal to one half of its normal value.

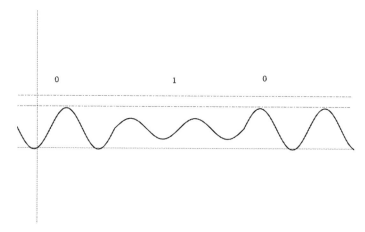

Figure 11: Amplitude modulation: the signal 010. To send the bit value 1, the amplitude is reduced for about one wavelength.

There are other kinds of modulation, in particular **frequency modulation** (FM) in which some other property of the carrier is changed as the signal goes out. In the case of FM, the frequency of the carrier is jiggled slightly to encode the data; not quite enough to stop the antenna from working well, but enough to get the data across the wire.

An additional encoding mechanism for sending data via radio waves is phase-based encoding. Think of the signal being transmitted as an undulating wave (a sinusoid function). Shifting the function slightly to the left or right corresponds to a phase shift, as seen in the illustration below. The maximum amount of shift that is possible is limited, since by shifting the signal to the left far enough, you get back to the same picture (assuming the signal is infinite in each direction).

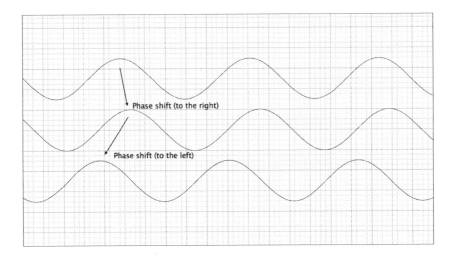

Figure 12: Signal phase. Sinusoidal signal shift right or left.

Slight shifts in the phase of the sine function, over limited chunks of time, can be used to encode digital data. In the sample cartoon below, the bit pattern 010 is encoded by shifting, or failing to shift, the signal over limited periods of time, first not at all (encoding a zero), then to the left (encoding a one), and then not at all again. This is illustrated below by also showing an unshifted reference signal for comparison.

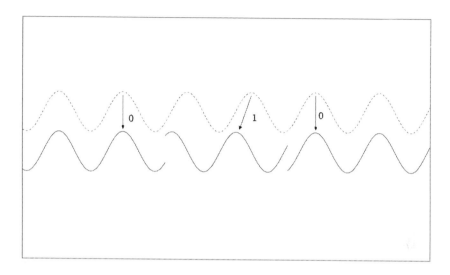

Figure 13: Encoding using phase modulation

The actual signal transmission methods used for digital television combines both amplitude-based signal transmission and phase-based transmission.

These ideas are useful in practice because there are up to three important frequencies that may be needed to set up and use a digital receiver, such as a satellite receiver, and it helps to know how they relate. These are: the carrier frequency that is actually received by the antenna, the intermediate frequency (IF) used to send the signal from a dish antenna to a receiver, and the actual channel frequency of the data. The IF is just a different carrier frequency that works well when carried across a wire, as opposed the higher frequency used to transmit signals through the air.

Signal transmission

We have seen that there are several ways to modulate a carrier frequency with a signal of interest to get data from one place to another. The actual signals being sent from one place to another are transmitted using physically continuous signals (amplitudes, phases, etc.), but they are interpreted digitally. Getting a digital signal transmitted depends on making a connection between specific physical values, such as amplitudes, and specific symbolic values, such as zero and one. In addition, since the transmission is rarely perfect, we need of way of deciding how to deal with borderline signals that are hard to interpret. For digital signal transmission, in particular, it is also important to have a strategy for correcting errors.

One of the dominant and elegant methods for sending digital television signals is called **Coded Orthogonal Frequency Division Multiplexing** (COFDM). It has several different variations based on modulating both the amplitude and phase of the transmitted signal at the same time. As a result, instead of sending only one bit at a time, COFDM is especially well suited to sending successive symbols made from an alphabet of more than just two possibilities; in other words sending multiple bits at a time.

The successive transmissions are referred to as **symbols**, each of which can take on one of the several values of our alphabet of possible values. For example if a symbol can take on 8 values, based on its strength, then receiving this single symbol provides 3 bits worth of data since it requires 3 bits to encode a number between zero and seven.

> Symbols are like digital bits, but they can take on more than just two values. In fact, bits are a special case of a two-valued symbol.

A common technique for doing this is "64 **QAM**," which also allows a single signal to be jointly carried by thousands of different carrier frequencies, with the hopes that some of these will be received relatively interference-free (there is a 2000 and an 8000 carrier version). QAM is a form of amplitude modulation (seen on page 38) and it stands for Quadrature Amplitude Modulation and used extensively for cable television transmission. 64-QAM refers to use of QAM modulation with symbols that take on one of 64 different values, meaning each symbol carries 6 bits of data.

Why is the picture quality of digital signals better than that of analog signals? One reason is that newer TV standards allow for larger number of pixels to be sent and displayed, but the more fundamental reason is that digital signals can be encoded in such a way that errors can be automatically detected and then (sometimes) corrected.

Automated error *detection* alone still leads to some sort of anomaly (like a briefly frozen picture), since knowing that an error is present in the transmitted signal is good, but something still has to be displayed on the screen. Error *correction* refers to not just detecting, but repairing errors, and implies that extra data has to be sent with the original signal, so that once an error is detected, there is enough extra information present to allow it to be fixed. For example, the simplest unrealistic case would be to send every video image in a television transmission twice, so that if an error is detected in a pixel, a correct value of the pixel could be obtained from the duplicate copy of the image. In practice, real error corrections methods are much more elegant and efficient (and complicated) than just sending the data twice. When error correction information in sent in advance, to be used just in case an error is detected, this is known as **Forward Error Correction** (FEC). FEC is commonly used with almost all digital communications, including video.

4. Switching from Analog to Digital

As digital television becomes more prevalent, many consumers will want to switch from home entertainment systems that handle only analog video to ones that accommodate digital video. In this chapter we provide a few practical suggestions on how to make this transition.

Does your TV (and other gear) do digital?

In general, most modern digital television systems, including HD and DVD players and satellite receivers, will output a high quality standard definition signal you can view on any television, even an older analog-only one. There are several different kinds of connector for each of both analog and digital signals, but most output devices support a basic selection. Likewise, new HDTV television sets always have input for traditional standard definition analog signals, and can display them on.

Moving to digital gets complicated due to two issues. The first of these is the desire to receive digital signals over the air, and the second is the desire to see the best possible picture and make the most of the video components you already have.

Most television sets made before 2007 will not receive over-the-air digital broadcasts based on ATSC (North America) or DVB-T. Only television sets labeled as either ATSC, DTV or DVB-T (in Europe) are likely to be able to receive any over-the-air digital broadcasts with their built in tuners. Sets labeled "digital tuner" or "digital receiver" will also work, but those labeled only "HDTV" may not actually have a suitable over-the-air tuner built in. Those labeled QAM will work with digital cable (under certain conditions: see Chapter 7).

If you want to watch high definition signals from Blu-Ray disks, HD satellite programming, or over-the-air broadcasts or cable, the story is more complicated. Almost every receiver for these kinds of programming can also output standard definition signals that can be used with older equipment. These standard definition outputs are pretty much always analog signals, and they may not actually reflect the resolution advantages of HD programming, although they will include the noise immunity of digital signals and thus provide nice clean images.

Most televisions made before 2006 probably will not handle true full-resolution digital high definition television (aside from the fact that the actual definition of resolution that defines HD is a bit vague, as discussed earlier). Even some of the digital HDTV television receivers sold in 2006 and before may not properly handle the encrypted signals that are currently output by essentially all modern devices for true HD programming. This is not because the televisions are not good enough, but because of more recent encryption requirements that were agreed on by manufacturers. As discussed in more detail the chapter covering Digital Rights Management and Copy Protection, today's high definition devices have intentionally crippled analog outputs, and will only provide full-resolution digital output to devices that include licensed encryption and digital rights management systems.

If your television has an HDMI or DVI input (described on page 250), and it was built after 2006, you can be pretty sure it can handle modern HD signals. If it lacks HDCP or DVI inputs, it definitely will not support full resolution digital HD signals from most of today's devices that support digital rights management (i.e. almost everything). That leaves two special cases: clear programming that doesn't use encryption, and older HD devices that don't support HDCP encryption.

A fair bit of HD programming is sent without any DRM, and some receivers will provide this in analog format at high resolution. If your analog TV set is fancy enough, it might be able to display digitally transmitted analog video in 720p format.

The really sad cases are those people who purchased expensive HD television sets before HDCP digital encryption was fully defined and mandated. In these cases, these devices are able to display full resolution output, but most modern devices will refuse to provide them with the high-resolution signals they require, because they might not comply with the DRM restrictions desired by the providers of the media. Devices that convert digital video signals to analog, known as transcoding the signal, do exist. HDCP encryption (see page 181), however, is meant to prevent this. Transcoding devices do appear on the market now and then, but usually violate various licensing agreements and thus are hard to find and may only be available intermittently or with uncertain reliability.

Receiving digital broadcasts over the air

Over the air television broadcasting was the primary way for home viewers to receive television for most of the 20th century. Analog television receivers built in 1940 could, in principle, still be used in 2008 (assuming the tubes were not burned out). With the advent of digital television, however, analog receivers will no longer be able to receive programming. As we have seen, digital television broadcasting over the air is based on one of two standards, ATSC or DVB-T, where ATSC is used in the USA, Canada and a few other places, and DVB-T is used in Europe and everywhere else.

Over the air reception of digital television requires 3 key components: an antenna, a tuner, and a display device. All three of these were once combined in a single unit for analog television. Digital broadcasts use the available radio frequencies in a different manner from analog broadcasting and require a different receiver, although in some cases the same old antennas can still be used.

The digital signal that is picked up by the receiver is also based on a totally different video system from the older analog NTSC system used in North America, or PAL/SECAM elsewhere. It is possible, however, to transform an incoming digital video signal into an analog one. This means that by using a small "transcoding" converter box, a signal from a modern digital receiver can be rendered suitable for display or recording on older analog equipment.

The end of terrestrial analog

The US congress and the FCC mandated that in February of 2009, all full-power terrestrial (over-the-air) analog television broadcasting in the USA was to cease, and be replaced by digital broadcasting (using ATSC). Initially, the date for the transition had been set even earlier than 2009, but that was found to be impractical. This type of transition to digital broadcasting was also legislated by the Canadian parliament (for the year 2011). Several countries in Europe have led this transition. For example, the Netherlands made a complete transition to digital terrestrial broadcasting in 2006, Finland became fully digital for over-the-air broadcasting in 2007, and the European Union as a whole has proposed that all member states do so by 2012. In China, the changeover is slated for 2015. Obviously, making this transition is easier in countries with a high standard of living since it means replacing or upgrading many receivers, as well as broadcast equipment.

In the USA, the legislation that required the transition to digital also required cable broadcasters to carry local broadcast stations, including in analog form, for subscribers. Satellite broadcasters have also been increasing their coverage of local broadcast channels for some time.

Clearly, analog and digital signals can both be broadcast at the same time, and this has been the case the word over for many years. Why require the analog signal to be terminated by law, as opposed to letting market forces play out? There are several justifications for legally requiring analog broadcasting to terminate.

The standard argument for digital over-the-air broadcasting is that broadcast spectrum is a scarce resource: there is only a limited range of frequencies available for over-the-air broadcasting, and these are highly regulated. Digital broadcasting makes more efficient use of these frequencies, and this represents a better use of the available spectrum. Thus, by shutting down analog broadcasting, the more efficient digital broadcasting system has space to grow. On the other hand, since broadcasting in most of the Western world is driven by market forces, why not let the demand for digital channels simply move the market forward, and let the spectrum space for analog simply be squeezed out by pricing? There are several motivating factors one can suggest. By legislating a switch to digital, the changeover is assured to be far more rapid and complete. Making the transition via legislation assures synchronization between different broadcasters and manufacturers, without which a switch to digital technologies might be very gradual. Changing to digital terrestrial technology also encourages a switch to HDTV display and recording technology as well, since televisions with build-in digital receivers are universally HDTV. The result of this is a huge demand for new electronics and a surge in economic activity, which may result in substantial profits and activity in some parts of the economy (albeit at a cost for consumers who are "forced" to switch over). Lastly, the creation and duplication of analog video recordings is very difficult to control, while digital rights management and copy protection for digital media are well developed (and can be expected to improve). As a result, there is probably substantial enthusiasm for a switchover a soon as possible from the content-creation industries.

Using an antenna

Most people who watched terrestrial analog television did so without using an outdoor antenna, counting only on the antenna attached to their television set. Although digital television uses the same frequencies as analog, and delivers a better picture if it delivers any picture at all, it is less tolerant of situations where there is a poor reception. Whereas an older television could be used with only an internal antenna to deliver a picture in all kinds of places, even if it might be fuzzy and full of static, a digital transmission will require a certain minimum signal quality and thus will often mandate an outdoor antenna.

This need for an outdoor antenna will probably come as a surprise, a disappointment, and even a shock to many who convert to digital. It my well be the best thing about digital from the point of view of subscription broadcasters, and specifically the satellite and cable television broadcasting industries since even more people will probably just give up on terrestrial broadcasting.

Adding just a digital receiver

For those who already have a standard definition television or home entertainment system, a stand-alone digital receiver can be used to pull in terrestrial digital broadcasting, and then produce a video output in some preferred format. If this output is analog, then it can be used with traditional analog recording and display devices such as a VCR, a video frame grabber, or a television set.

The converter box

A "converter box" is a self-contained receiver for ATSC or DVB-T that pulls in digital television signals over the air, and produces a regular television output. The US government decided to subsidize the sale of simple converter boxes with basic analog outputs, in order to ease the transition to digital broadcasting. The rules that were passed regarding these subsidies stipulated that the converter must be *incapable* of producing anything more than a basic analog output using composite video. Receiver boxes with *both* analog and digital output, or with higher-quality component analog output, or other nice features certainly exist, and they can be connected to various kinds of televisions, but they were deemed *ineligible* for the US subsidy. The key practical for a consumer issue is to assure that any converter box has outputs that match those on the television system with which it is to be used.

Upgrading only the TV

Instead of buying a stand-alone receiver (converter box), many viewers choose to purchase a new television. In conjunction with the move to digital broadcasting, many consumers decide to update their television set (or monitor) to one that supports HDTV resolution. Almost every HDTV display accepts both HDTV as well as standard definition inputs, so backwards compatibility is rarely a problem. The two concerns to beware of in that regard are: (1) if you are considering an older used display (pre-2007) then be sure the digital input supports HDCP encryption over HDMI cables (page 181), which is required by most modern video sources, and (2) make sure the standard definition inputs match the cable types on your existing gear. In particular, the best quality analog signals can be obtained using component video cables (page 247), and if your other gear, like a DVD player, supports it then the television should also.

As we have seen, HDTV uses an aspect ratio of 16:9 making for a wider screen than standard definition, and it also means increased resolution. Exactly how much resolution qualifies as HDTV (as opposed to "genuine DTV", "real HDTV" and "true HDTV") is open to discussion, but HDTV resolution television sets typically support a resolution of at least one of 720p, 1080i, or 1080p (where 1080p is the best and most costly choice; see the discussion starting on page 22).

In addition to the choice of resolution, the other big issues are the physical size of the screen and the display technology. As far as the physical size goes, bigger is usually better except for the not-so-minor issues of affording the price, and physically fitting it into the room. Don't forget to consider the fact that larger displays call for longer viewing distances, so if you only plan to sit 3 feet (one meter) from the television, a television with a screen 65 inches (165 cm) across may not be suitable. Sitting too close will expose faults in the display itself, and will lead to a worse experience perceptually. In addition, especially with standard definition recorded content, it may look less pleasing if it is too big relative to the viewing distance. The typical rule of thumb is to sit no closer than a distance 1.5 times the diagonal length of the screen. These values work out as follows:

Diagonal screen measurement	Minimum viewing distance	Max viewing distance
26" (65 cm)	3.3 ft (1 m)	6.5 ft (2 m)
30" (75 cm)	3.8 ft (1.15 m)	7.6 ft (2.3 m)
34" (85 cm)	4.3 ft (1.3 m)	8.5 ft (2.6 m)
42" (105 cm)	5.3 ft (1.6 m)	10.5 ft (3.2 m)
50" (125 cm)	6.3 ft (1.9 m)	12.5 ft (3.8 m)
60" (150 cm)	7.5 ft (2.3 m)	15 ft (4.6 m)
65" (165 cm)	8.1 ft (2.5 m)	16.2 ft (4.9 m)

Many televisions also include video processing technology that cleans up a picture, which can be especially important for watching standard definition content on a large screen. The reaction to this kind of technology tends to be quite subjective, and the types of processing vary significantly by manufacturer, so each user must make their own assessment.

Old recordings, new TV

Almost all digital display devices also accept analog inputs in various forms. Thus, if you have playback devices for analog video, such as DVD players, or a VCR, connecting it to a digital television and directly viewing the result should be straightforward. The only catch is to make sure the cables between old and new devices match up.

Often, however, old media can look unsatisfactory on a new television display. Assuming there is no physical problem, like bad wiring, this can happen for two principal reasons.

The simplest of cause of unsatisfactory display of standard definition content is that modern digital displays tend to be much larger than those that were used for standard definition television, as well as having higher brightness and contrast. As a result, older content may not look so good. Making sure the contrast settings on the display are not too high can help. Many displays include noise reduction algorithms that are supposed to help too. The simplest and most commonplace of these simply uniformly blur the image slightly.

A second more exotic cause of problems is the performance of the noise reduction process itself. In some cases, and in the opinion of some viewers, the automated processing that attempts to make SD content look like HD content actually makes it look worse. In most cases, it can be turned off.

For the discerning viewer, there are devices that will take SD content and apply much more sophisticated processing to upscale it to HD resolutions. These devices, known variously as scalers, upscalers, or upconverters tend to be costly and almost arbitrarily exotic. The key to such devices is to select what kind of sharper higher resolution pixels can be used as replacements for each group of low resolution pixels.

5. Satellite TV Overview

The chapter provides a brief outline of how satellite television works, from the point of view of the ordinary home consumer. More subtle, uncommon or exotic features are covered in later chapters.

A satellite television system can be broken down into three very different parts: the program transmitting source, the satellite in space, and the receiving end. The transmitting source involves the preparation and selection of signals, including inserting advertising and encryption. The satellite in space is owned or leased by the transmitting company and deals with relaying signals to various parts of the earth. The receiving end is located in each consumer's home.

In general the companies that transmit the satellite signals do not actually create the content, and so they need to purchase program content which is often sent to them either via another satellite "feed," or over land-based cables. The receiving end is the **integrated receiver decoder** (**IRD**) (i.e. receiver) owned by the home user.

Satellite television is almost always based on a satellite that is fixed in space relative to the earth's surface. This allows the transmitting antenna and receiving antennas to be permanently oriented to point at the satellite. Satellites are usually identified by a name and by the latitude around the earth's surface (which always includes a letter for East or West latitude), for example the Eutelsat Hotbird 6 satellite is located at 13° E (over Europe) and the Dish Network EchoStar VIII satellite is at 110° W (over North America).

The most common satellites in use for television today in the Americas and Europe are based on so-called Ku (pronounced Kay-you) band signals, possibly with a layer of digital encoding that is known as DBS or DSS. These are signals transmitted via satellite and received by the home user on a moderate-sized dish-shaped antenna which must be located outdoors. Other signal frequencies, to be discussed later, are known as C-band and Ka-band.

The broadcasting process involves traditional video-handling systems that prepare and select content that is finally sent up in space via the use of a huge transmitting satellite dish. The place where this takes place is referred to as an **uplink facility**. The key steps there involve:

- receiving content from the various content providers (i.e. television networks),

- recording them at least temporarily,

- inserting advertising into the shows,

- combining the shows together to make individual channels,

- digitizing and combining channels together into a data stream to be sent up to the satellite, typically on different radio frequencies,

- and encoding the data to provide data security.

Finally, the data are transmitted to the satellite in a step called the **uplink**. The transmitting dish has a typical size of 30 to 40 feet (10 to 13 meters) and a transmitting power of approximately roughly 100 Watts per transponder (this can vary from 16 Watts for old C-band technology, to 100 Watts for Ku-band signals, to 240 Watts for digital DBS signals). To put this in perspective, note that 100 Watts is about the same at a typical incandescent reading lamp bulb, and this signal is distributed

Quick summary:

Most digital home video is known as DBS and arrives in the Ku band.

over a huge area by the satellite.

Figure 14: Satellite transmitter dishes. Photo by Timo Newton-Syms provided under Creative Commons attribution license.

Various satellites are used for television transmission. The job of the satellite is fairly simple, it receives the signal from a transmitter located somewhere on earth (the uplink), and re-transmits the signal back down to earth to cover a wide area (this is called the **downlink**). The satellite can thus be regarded as a big mirror floating in space that just reflects back the transmitted signal; think of how mirrored disco-ball at a nightclub can take the light of a single spotlight and return it to many different points.

In reality, however, the satellite does not actually just reflect the signal, it captures it and transmits it again in selected directions. In fact, it retransmits it with a different frequency from what it arrived with, and shapes the outgoing beam as it does so. Any satellite can directly "see" almost half the earth's surface and thus cover all that space with its broadcasting. Doing so is known as a **global** transmission (see Fig. 15). In most cases though, sending out enough signal energy to adequately cover half the earth is not efficient since much of the surface "seen" by a satellite may not contain potential viewers (either due to national boundaries, or because there is ocean there). As a result, most satellites have specially engineered broadcasting antennas that focus the outgoing beam to some extent and conserve energy by only beaming the signal to a restricted region. This practice is sometimes called using a **spot beam**. The extent of a beam can be as narrow as a couple of states of the USA (for example the satellite Galaxy-28 at 89°W has many small spot beams covering the USA, one of which is confined almost exclusively to the state of Texas), or as wide as North and South America combined (for example the transmissions of Intelsat-9 at 302°E).

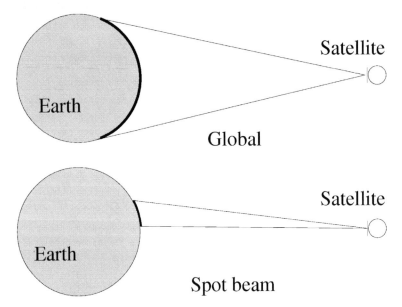

Figure 15: Global broadcast versus Spot Beam. The shaded portion of the earth is the part of the world able to receive a broadcast from a satellite. With the Spot Beam, the transmitted energy is restricted to, and concentrated in, a smaller region.

The last stage is the reception process in your home. Since the satellite signal comes from very far off, and the satellite itself only has a limited amount of energy, the signal that comes from a satellite is quite weak (as compared from a terrestrial TV or radio station). As a result, the antenna receiving the satellite signal needs to be large (compared to a pocket radio antenna), focused to collect the signal, and placed outdoors: this is the satellite dish antenna. This antenna receives a wide range of radio frequency signals, and focuses them at a single spot. At this focus we have the LNB, another antenna-like device that selects a small set of the available signals and transforms them into a something that can be sent up a coaxial cable to a receiver inside the home. This receiver extracts a single channel from the group of signals on the cable, decrypts it if necessary, separates the audio and video parts of the signal, and produces a television or audio program you can enjoy.

Most people today use a dish antenna to watch digitally encoded programs, and the dish is fixed and immobile. Depending on weather, ambition, available space and other factors, some people use large C-band dishes that can be aimed in different directions as desired, using automated motor systems.

An even more exotic class of systems are those based on **Very Small Aperture Terminals** (VSAT). These are systems that are bi-directional: the endpoint (such as a home or business user) can both receive data via the satellite antenna, as well as re-transmit data back to the sender. The term VSAT refers to bidirectional systems where the remote (e.g. home) earth station has a small antenna, typically 30 to 48 inches (75 cm to 1.2 m) in diameter. VSAT terminals are used for Internet connections, business data processing and scientific processing at remote locations.

A whole extra hidden layer of complexity related to the management of the telecommunication satellites. This involves extra signals sent to the satellites, the monitoring of their position and behavior, and adjustments to their operations.

The company operating the satellite is often the same as the company providing the television or data services, but not necessarily. Satellite operating companies rent satellite time in different ways. You can rent a short slot on a telecommunications satellite for a specific broadcast, or you can lease an entire transponder on a long-term basis. For example, the monthly rate for a transponder on a member the TELSTAR satellite family is on the order of between $100,000 and $250,000 dollars per month, depending on the bandwidth, power and other features (since TELSTAR has different kinds of transponders).

Up in space

The operation of the satellite itself is not of much practical importance for most users, but it provides useful insight on some of the peculiarities of how satellite TV and radio operate. Thus, we'll discuss the so-called "space segment" of the satellite broadcast system in this section.

The job of the satellite, as we have already seen, is to receive an uplink signal from a single broadcast station and return it to earth as a downlink signal over a wide region of the planet. Even if the uplink signal is focused as much as possible, it will still spread out and weaken as it moves away from the transmitter. Since the uplink signal will be quite weak by the time it arrives at the satellite, this requires that the satellite include an appropriate amplifier to boost the power of the outgoing signal. In addition, since the uplink and downlink signals will be streaming continuously, and hence simultaneously, they should be at different radio frequencies to avoid interference with one another, and thus the frequency of the downlink signal needs to be shifted from what is received. Finally, to provide more power, the downlink signal is subdivided by an **input multiplexer** into discrete frequency bands and these are handled by physically distinct amplifiers known as **transponders**. More generally, the word transponder is used to refer to all the components used to process a single frequency band with in the satellite. An amplification stage per band (i.e. per transponder) can use techniques such as a traveling wave-tube amplifier (**TWTA**) for Ku band signals, or transistor-based solid state amplifiers (**TSSPA**) for lower-power L and C-band signals.

After amplification the signals are recombined into one or more output streams by the **output multiplexer** for transmission. Most satellites have between 16 and 32 transponders with 24 being a typical number for a DVB satellite. Each transponder has a transmitting power of approximately 100 Watts, although this varies widely according to the type of satellite. Older C-band technology satellites use transponder power levels in the range of 16 to 32 Watts each, which is part of the reason that old C-band dishes have to be comparatively large. Satellites used for Ku-band transmissions, on the other hand, transmit at power levels of roughly 50 to 100 Watts per transponder. The DBS signals used for digital satellite broadcasts, although technically part of the Ku frequency band, are transmitted with up to about 250 Watts of power per transponder, which explains why these dishes can be so small (another part of the explanation is the efficient error-correction that is possible with digital signals). Transponders used for digital television have a bandwidth of 27 to 74 Megahertz, which is a measure of how much data they can transmit per second, and a typical figure is 36 MHz which provides roughly 38 Megabits/second of data.

In some cases, the broadcaster prefers to send different output signals to different areas, in the form of a "spot beam" signal as we saw above. In addition to make better use of energy, this is used, for example, to provide different local stations to different regions and is achieved by combining specific signals in different output multiplexers. Each of these output multiplexers is sent to a different transmitting antenna, which will be responsible for the reception over some portion of the planet that it blankets with a sufficiently strong signal. This part of the planet that gets signal from a given transponder is referred to as the **footprint** of the antenna (see the figure below where this is illustrated).

A more formal way to treat the issue of how the transponder signal varies with location is via a concept called **Effective Isotropic Radiated Power** (EIRP). This is a numerical estimate of the net radiated power in a given direction from the antenna at a given frequency, that is the power coming out of the satellite in a given direction. The term is also used (somewhat imprecisely) to refer to the received power on the ground. Technically is it equal to the product of the power in Watts to a transmitting transponder and the antenna gain in a given direction, but in practice for home users it is a map of signal strength as a function of geographic location on the ground (i.e. a quantitative description of the footprint). The larger the EIRP, the better the signal and the smaller the dish needed to receive the signal.

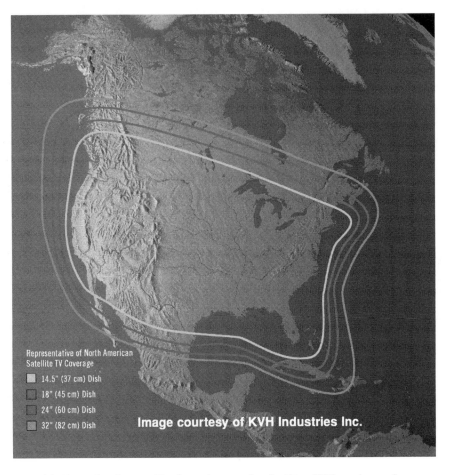

Figure 16: Example of a satellite footprint map for the DirecTV broadcaster's signal. Such a map varies with changes in system configuration. Image courtesy of KVH Industries Inc.

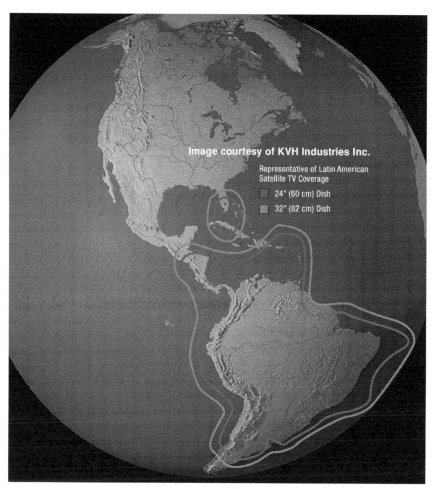

Figure 17: Example of a satellite footprint map for one of the Galaxy satellites (such a map varies with changes in system configuration and this is an example only). Image courtesy of KVH Industries Inc.

Orbits, positions and slots

The satellites used for television broadcasts and most Internet services are fixed in the sky with respect to an observer on earth. That is, they are in geostationary orbit (known as GSO)—stationary with respect to the earth's surface. How can this be maintained, you might ask? The satellite cannot simply float in a fixed position without falling down due to gravity. In reality, the satellite is *not* in a fixed position with respect to the *center of the earth*. It is in a genuine orbit, rotating around the planet constantly to stay up in space. Since the earth is rotating too, it is possible to remain in a fixed position with respect to *a point on the surface* (which is also rotating). To do this, the satellite needs have an orbital velocity that is exactly synchronized with the rotation of the planet. For this reason, such orbits are also geosynchronous (GEO): *geo* for earth, and *synchronous* for the synchronization with the rotation of the surface. Not all geosynchronous orbits are stationary, however. Even when the orbit is synchronized with the earth's rotation, it also needs to be directly above the equator to be geo*stationary* (otherwise in order to circle about the center of the planet it needs to also move North and South, returning above the same point on the surface only intermittently).

The satellites used for the Global Positioning System (GPS) are in semi-synchronous orbits. This means that their rotational period (the time for a full orbit) is 12 hours instead of 24 hours, and thus they do not remain at fixed positions relative to the surface. For satellite television this would be very problematic since the dish needs to be aimed at the satellite, and if the dish had to track the satellite across the sky the entire system would be much more costly, error prone, and less reliable.

In order to maintain a geostationary orbit, a satellite must not only remain over the equator of the earth, but it also needs to have an orbit with the correct radius (and thus be at the correct height over the surface). This height follows from the relationship between gravity and centripetal acceleration, the force you feel when you swing an object on a string around your head. To put it another way, the orbital radius and the rotational period are directly related to each other. A geostationary orbit about the earth needs to have an orbital period of 23 hours and 56 minutes (which is the actual length of exactly one day, and thus one revolution of the Earth itself). This period corresponds to a height of 35,600 km (22,120 miles) above the earth's surface, or an orbital radius of 42,000 km (26,097 miles) relative to the center of the earth. The key practical observation here is that a geostationary orbit can only be achieved at a specific height directly above the equator. This combination of restrictions with respect to latitude and height defines a ring around the earth where all the geostationary satellites need to live. As a result, geostationary satellites all need to live in a circumscribed band around the equator that is sometimes known as the Clarke Belt, after the science fiction writer Arthur C Clarke who was the first to propose the practical use of geostationary orbits. This ring is getting increasingly crowded. Before geostationary satellites come "on line," and after the end of their useful lifetime, they typically live in nearby orbits that are geosynchronous, but not geostationary, and thus they wave up and down in longitude as they orbit while remaining locked in latitude.

In the case of satellites broadcasting similar services, such as those used for Ku-band television broadcasts, the satellites also need to be somewhat separated from one another to avoid radio interference as well as the risk of collision Thus, we have a ring around the earth chopped into **orbital slots** that the satellites can sit in. Sometimes more than one satellite can share a single slot, but only under special conditions (such as when they are owned and operated by a single company in a coordinated manner).

The placement of all satellites in a single ring is why all satellites used for television can be described with just one orbital angle (like 110°W). Just the longitude alone provides the position of the slot around the ring. If you know the longitude, then the latitude is given by the equator, and the height above the surface it fixed, and thus the satellite is precisely located in 3-dimensional space. To find it from a fixed observation point on the ground is another matter, however, and the "look at" direction depends on your position on the earth's surface. To do the pointing, you need two angles that depend of where on the Earth you are looking from (see figure below).

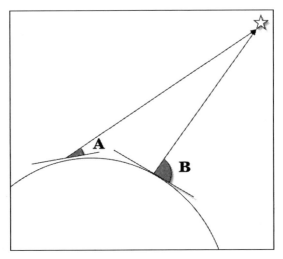

Figure 18: From different positions, the direction to a fixed object in the sky shown as (A) and (B) relative to the ground, may be different.

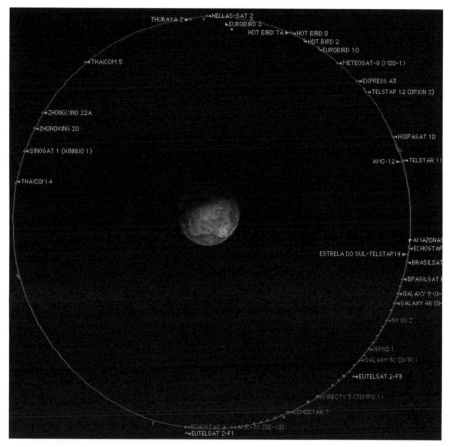

Figure 19: Picture illustrating the positions of various commercial geostationary telecommunication satellites about the earth.

Based on the height of a geostationary orbit above the earth's surface, about 40% of the earth has a direct line of sight to any single satellite. This doesn't always mean that people in all of that 40% can receive the broadcasts from that satellite, since this also depends on signals strength and antenna properties as well, but 40% of the earth has at least a chance to see each geostationary satellite in the best case, if the satellite is engineered correctly.

As an aside, it might be worth noting that even with an ideal geostationary orbit, a satellite will tend to drift slightly due to various factors including the solar wind (radiation pressure caused by particles coming from the sun), residual atmosphere, or the effect of collisions with miniscule particles. As a result, a telecommunications satellite needs to occasionally make small altitude and position adjustments using thrusters that burn stored fuel. Since this fuel will gradually be used up while the satellite is in operation, the satellite will have a finite usable lifetime before all its fuel is used up. Eventually after a decade or two, every satellite has to be moved into a higher supersynchronous disposal orbit.

Some functioning satellites occupy orbits that are almost geostationary, but which wobble slightly, often due to lack of fuel. Such satellites are referred to as being in **Inclined Orbit Operation** (IOO).

Receiving the signal

The last leg of the broadcasting process is the receipt and display of the signal at home. This is accomplished by a combination of an antenna to pull in an otherwise weak signal, a tuner to select a frequency of interest from the mix of radio-frequency energies tapped by the antenna, and digital decoding circuitry to pull out and display the digitally encoded packets that make up a program of interest. For digital programming (DVB or ATSC), the tuning and digital processing are typically handled by a single self-contained unit called an **integrated receiver-decoder (IRD)**, or else by a computer card that accomplishes the equivalent functions. The antenna is an outdoor mounted dish. It has become

> **Quick summary**:
>
> Direct-to-home satellite broadcasts are received on an receiver known as an **IRD**, which connects to the television set. The IRD sometimes includes a video recorder.

commonplace for the IRD to include a digital video recorder as well (a DVR). We will discuss both on these in more detail later on.

Satellite radio

The technical name for satellite radio is Digital Audio Radio Service (DARS) and allows subscribers to tune in a large number of radio stations and provides them over a large geographic area. Like satellite television, satellite radio works by sending a signal up to a satellite from a base station on earth and then having the satellite beam in down over a geographically large "footprint" on earth. In addition, like digital satellite television, and unlike older analog radio and television, the receiver doesn't tune in just a single channel, but a data stream made of a collection of channels. The radio program making it up is divided into packets of data and these are piled together like the cars of a train. The program of interest selected by the receiver is assembled from the various packets to make up a show, and of course this all works because the amount of data that can be sent rapidly via the satellite is very large. Superficially satellite radio is just like digital satellite television, but on closer inspection there are a number of subtle differences.

Satellite radio is meant to be a portable technology that can go in a car, but is means it needs to provide reception even when there may not be a good direct view of a specific satellite. In big cities, the direct broadcasts from the satellite are often supplemented by broadcasts from additional low-power terrestrial retransmission stations to avoid dropouts or interference from tall buildings (between which the spaces are sometimes referred to as **urban canyons**).

Satellite radio in the USA uses the S-band for broadcast to consumer antennas, while elsewhere in the world the L-band is used. Since satellite radio is intended for use while moving, the antenna cannot be permanently aimed directly at the transmitter. A non-directional antenna can be used because satellite radio is broadcast at higher power levels than satellite television. In addition, each broadcaster is assigned a distinct frequency range, so that there is less need to selectively tune into only the transmissions from a single direction: there are no competing satellites at the same frequency that need to be ignored. A side effect of using non-directional receivers is that the satellites do not have to be in geostationary orbit, and some broadcasters have selected to use geosynchronous orbits instead.

The actual data rate a receiver delivers is around 4.4 MBits/sec of digital data. This is subdivided among various radio channels so that about 100 compressed programs can be accommodated. Since the necessary data rate per channel is much lower than television, it is practical to store several seconds worth in memory. As a result, another trick that can provide immunity to brief signal interruptions is to buffer a small amount of programming so that if the signal is interrupted, the receiver has a cache saved data it can continue to play for a few seconds until the reception improves.

6. Terrestrial Broadcast Overview

The terrestrial broadcast system refers to the broadcast of television and radio from ground stations directly to homes or automobiles. This is the traditional approach to broadcasting that was developed at the start of the 20th century by Marconi. Terrestrial broadcast is often referred to as "over the air" broadcast, or OTA. Over the air signals can be analog or digital, with the move to digital gaining ground and being legislatively mandated in one country after another. In the USA, legislation was passed to move almost all terrestrial broadcasting to digital formats as of Feb. 2009, making all older television sets obsolete (as discussed in Chapter 4).

The analog broadcast networks

The discussion of analog broadcast networks is an ode to a vanishing technology. Analog television broadcasting would probably persist indefinitely, but it was mandated out of existence in one country after another. In the USA, its use subsequent to Feb. 2009 is reserved for special-purpose applications such as low power regional broadcasting. The primary reason for the mandated switch to digital TV is that digital broadcasting makes more efficient use of the available bandwidth. Thus, if analog broadcasting is replaced by digital, the spectrum space is made available for other uses such as cellular phones or additional channels. More cynical explanations include the fact that forced adoption of digital technology closes the what the rights holding industries called the analog hole, meaning the ability of home viewers to record an event and distribute programming they received.

Analog broadcasting uses two different frequency bands, both of which are used for digital television broadcasting as well. These bands have typically colorful names used for radio spectra: the Very High Frequency Band (VHF) and the Ultra High Frequency Band (UHF). Transmission on both bands uses similar principles, with video being sent on one frequency and audio being transmitted on an FM "sub-band" just at the low end of the range that is used for traditional FM radio transmission. The VFH band is used to broadcast channels 2 through to 13. It is subdivided into two sub-ranges, with the low part dedicated to analog channels 2 through 6 (54 to 88 Megahertz) and high part being used for channels 7 through 13 (174 to 216 MHz). The UHF band is used for channels 14 through 69 (470 to 806 MHz).

Digital terrestrial overview

Television stations the world over are eyeing the transition from convention analog television to digital TV. This step has been talked about for many years, but the transition is difficult since simultaneous broadcast using both types of signals is costly, and because so many traditional analog systems are already in place. Nevertheless, in the United States congress passed a law forcing a wholesale transition to digital television at the end of the first decade of the 21st century. In Europe, the transition seems to be talking place on a more incremental consumer-driven basis.

> **Quick Summary:**
>
> Digital television broadcasting uses **ATSC** in North America, and **DVB-T** in Europe. They're almost the same thing, but require different receivers.

No matter how the transition occurs, there are two natural pathways: via a new digital television, or via an intermediate set-top box that converts the incoming digital signal to an analog signal compatible with an existing "legacy" television. In either case, the television or set-top box contains a digital receiver. In fact, the operation of a terrestrial (over-the-air) digital television system is very similar to a digital satellite system. The only difference is, of course, that with terrestrial broadcasting the signal is sent directly to the home viewer from the equivalent of the ground station (i.e. the transmitter), and the radio frequency encoding and decoding is different. This difference in encoding is primarily due to the need to cope with different kinds of interference issues on the ground, such as the fact that the signal can more readily bounce around between buildings. As a result, a different encoding system is used. Two or three systems have been developed, one for use in the United States (ATSC) and a few other countries, another international standard in use in most of the rest of the world (DVT-T), and one restricted to Japan and Brazil called (ISDB-T). As the level of detail we working at, ISDB and DVB are almost indistinguishable at a software level. The biggest difference between DVB-T and ATSC is the way the actual radio signal is transmitted, as opposed to the way the digital data is encoded, which is quite similar between all three digital TV systems.

The DVT-T transmission system that is used to send the signal from the station to the home is based on a radio-coding technique called Coded Orthogonal Frequency Division Multiplexing (COFDM) that we saw earlier. COFDM is used for both terrestrial DVB transmission as well as ISDB-T. It has several different variations based on modulating both the amplitude and phase of the transmitted signal. It allows a single signal to be jointly carried by thousands of different carrier frequencies, with the hopes that some of these will be received relatively interference-free (there is both a version using about 2000 different frequencies, and a version using 8000 frequencies). In addition, some of these frequencies are "pilot" signals that are just used to measure the amount of corruption they experienced, to help calibrate the system. The pilot signal serves as a kind of test run of the transmission medium, to help determine how well it works on a moment-by-moment basis. In addition, some of these separate components of the total signal are slightly delayed relative to one another (by very small durations), again increasing the chance that the data gets through even if all the carriers are interrupted very briefly (for example due to a low-flying airplane). Since such a system can tolerate, in fact it can exploit, differences in arrival time for the signals, an ingenious mechanism to improve transmission quality has been developed. This is to use multiple transmitters sending the same signal. For analog TV, these transmitters would interfere with each other and cause multipath interference (ghosting, discussed on page 145). Since the different distances to the transmitters from a particular receiver would lead to slightly different arrival times for the signals, but with digital COFDM these multiple sources just lead to different time delays and if one pathway is better then the other it allows for improved signal quality.

The first step a receiver for these signals needs to accomplish is tuning in a selected frequency that corresponds to one or more channels (channels can share frequencies in the digital world). The tuner, just as with analog TV, selects a frequency of interest from the incoming radio signals. A demodulator (for Coded Orthogonal Frequency Division Multiplexing (COFDM) or 8-VSB (for ATSC) is used to separate the actual signal from the carrier frequency. In North America, a modified version of the 8-VSB protocol is often used which adds additional error correction for better reception, and produces the **Enhanced** 8-VSB protocol, or E8-VSB. This signal contains the data of interest, but it is still in the form of continuously varying waveforms. The selected frequency then needs to be digitally processed to and digitized to recover the encoded packets that make up the data. Each single frequency has enough capacity to hold about five actual television channels.

A further embellishment to the processing of sending terrestrial signals is to send two different signals at the same time, a "high priority" (HP) one and a "low priority" (LP) one. The idea is for the HP signal to be a lower quality version of the signal sent in a lower bandwidth and less demanding manner. In the case of poor signal conditions, a receiver might be able to receive only the lower-quality HP signal, but at least it can display the program being sent in some form. When conditions are better, the LP signal is used to provide a better quality experience.

Terrestrial broadcasting using DVT-T can be extended to mobile devices by adopting the extensions described in the **DVB-H** protocol for mobile devices. In general this appears to be a straightforward extension. In the case of countries unfortunate enough to be using ATSC, the extension to mobile devices is more problematic. The modulation scheme used by

ATSC (8-VSB) is not well suited to mobile use and the transition to mobile versions of ATSC, although possible, is not as straightforward.

Terrestrial broadcasting

Terrestrial broadcasting stations that transmit the signals need to be almost directly visible from a receiver to the signal to be cleanly obtained. While the signal has a limited ability to pass around obstacles, this is only works for obstacles of limited size. In general, this means the receiving antenna needs to be pointed at the transmitter unless to local signal strength is very good. In Europe, where COFDM is used to transmit DVB-T signals, several different transmitting stations are usually used for the same broadcaster. In that case, accurate pointing of the antenna towards a specific transmitting antenna is much less important. Some of the issues related to using an antenna are discussing in the chapter dealing with antennas.

Program and channel Information

Digital terrestrial transmission in North America uses ATSC, which in addition to video and audio streams for each channel, includes a data stream called the **Program and System Information Protocol** (PSIP). This is supplementary information (sometimes known as metadata) that describes the programming being sent. The PSIP data plays the same role in ATSC broadcasts that the Service Information (SI) packets play in broadcasting based on the related DVB standard. Most importantly the PSIP identifies the content of different groups of data packets, the channel that is currently being received, and it provides information that specifies how a receiver can correctly cobine the streams to display the program.

The PSIP provides also descriptive information on the channel, supports delivery of the electronic program guide (EPG), delivers content advisory information, announces captioning services, and includes an accurate time-of-day clock. All these components can also be provided in the SI packets of a DVB-T broadcast.

The way service information packets are used to connect the various types of data in a broadcast is presented in greater detail in the chapter called "Digital Data Encoding: DVB."

7. Cable Television Overview

Cable television is, in some ways, the simpleton cousin of satellite and terrestrial broadcast television. With cable TV (sometimes knows as CATV, based on the obsolete phrase "Community Antenna TV") the program material is delivered to the home over wires, and so there is less need for fancy error correction protocols and schemes to deal with signals that may bounce off buildings when broadcast through the air. The technology has been around since 1948, making it much older than satellite broadcasting. While there are thousands of different cable systems in the USA alone, many of them are joined together in "interconnects" that allow them to share programming and advertising.

Much of the programming on cable TV comes from either satellite or over the air broadcast sources, although the satellite feeds may not be the same ones that most consumers have access to. Because wires are used all the way from the station to the home, it is suitable mainly for densely populated areas when the cables runs between amplifiers will not have to be too long. In addition, since transmission along wires can be very efficient, the possible bandwidth for cable TV (i.e. the number of stations or other data) can be very large. Finally, since a wire can carry signals away from the home as well as to it, cable based television has the best potential for simple interactive television services or feedback from the home user.

Analog cable TV programming is transmitted used the same types of technology as analog terrestrial broadcasting (although on a wider selection of channel frequencies). It can be directly received by all but the oldest televisions.

Digital cable television is transmitted used digital packets, just like terrestrial broadcast and satellite transmission. The packet format is either DVT-T or ATSC (recall that ATSC is a North American variation on the DVB standard). These digital packets are encoded as radio signals using Quadrature Amplitude Modulation (QAM, and more specifically 64-QAM), as discussed in the previous section of digital terrestrial broadcasting (page 42).

Many televisions are sold with tuners for both terrestrial over-the-air broadcasts, as well as cable. In North America these sets are usually labeled as ATSC/QAM, meaning that there is tuner for terrestrial ATSC was well as for unencrypted HD video using QAM over cable. In general, local broadcasts stations, for example, for available as unencrypted cable signals as well. These signals can usually be received by suitable QAM-capable televisions, even for those who only have analog cable subscriptions.

Digital cable TV based on DVB/ATSC uses digitally encoded data packets to transmit programming, and these packets are encoded using radio frequency signals. Internet television (IPTV), on the other hand, is an emerging technology that "wraps" the television data packets inside another transmission protocol used for generic data transfer over the Internet (i.e. TCP). Thus, one kind of container is put inside the other: this is like packaging the special purpose containers used within a company into generic cardboard boxes for so they can be sent via the postal service. In such a case, the underlying difference between cable TV and IPTV may be hard to discern for a consumer: in either case a wire from outside most into a "set top box" that is connected to the TV. We can expect to see such IPTV-based services to be increasing used for TV delivery in a way that makes them almost indistinguishable from cable TV.

The cable TV headend

To receive and collect all the different kinds of television material for transmission to cable TV viewers, a substantial facility is needed. This site is referred to as the *headend*, meaning it is the head of where all the cable TV material is sourced from. The facility almost always includes several large FSS satellite dishes and receivers, and antennas for the reception and retransmission of local broadcasting. Programs can also arrive in other forms included recorded rebroadcasts of previously received material (called time shifting), receipt of material via fiber-optic cable, or other media.

Instead of having a multitude of receiving systems and satellite dishes, some companies, such as the American Comcast cable company, broadcast using what they refer to as a **headend in the sky**. This is a small set of satellites that transmit a very large number of stations, thus providing most of the material for their ground-based headends.

Master antenna TV (MATV)

To receive and rebroadcast local terrestrial television, the cable TV headend facility typically uses one antenna for each individual station to be received. By using a separate receiving antenna for each station, the antenna can be optimally tuned and potentially combined with a frequency selective amplifier. This provides the best possible reception for each station. To produce the best possible signal for retransmission via cable, a collection of such channel-specific antennas are often combined into one structure. This collection of antennas is called the Master Antenna Television (MATV) tower (Fig. 20). These usually have a combination of UHF and VHF antennas.

The selection of channels received by the cable station is then combined to produce a single feed that is transmitted to viewers. In the case of analog cable, the different channels are assigned to different frequencies to allow them to be sent together (a technique called frequency division multiplexing). In the case of digital cable, they are separated into elementary program streams that are packetized and combined according to the MPEG protocol.

Figure 20: Cable TV master antenna. Round antennas are for UHF, VHF antenna can be seen in lower left. Photo courtesy of Andy Lamarre.

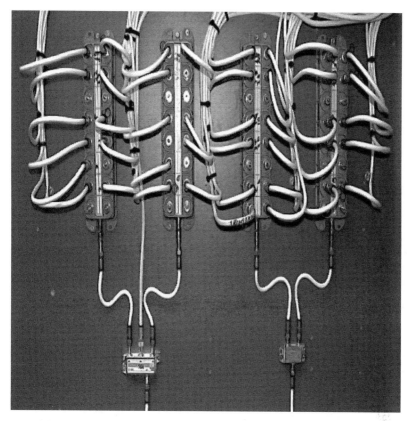

*Figure 21: Coaxial cables at a cable TV station for combining multiple channels.
Photo courtesy of Andy Lamarre.*

Coaxial and fiber networks

Cable television is delivered using coaxial cable that connects one or more devices in the home to a local distribution feed wire. The signal pulled down from the MATV antenna is amplified, shifted onto a new frequency (so that in can be combined with other channels) and then retransmitted to home viewers. This retransmission occurs using a city-wide cable wiring grid that is usually owned by the cable TV company.

Some areas are served directly by fiber-optic cable. A notable example is a service called FiOS offered in the United States, which delivers data using fiber optic cable right into a box installed in the home. This is unusual since fiber-optic cables are usually used up until a multi-home junction box, with the "last mile" to individual homes served by lower capacity coaxial cable (which is more than sufficient for almost any home user). Programming delivered using direct-to-home fiber-optic cable can be fed into a set top box like traditional cable TV, but the underlying transmission methods are actually closer to that used for internet-based television (IPTV).

SMATV: satellite/cable hybrids

Satellite Master Antenna Television (SMATV) refers to the aggregation and combination of multiple satellite channels into a cable TV output feed. A headend facility often includes this functionality, but as the required electronics become more compact and efficient, it has become possible for a single apartment building or hotel to operate their own in-house system. SMATV usually refers to fairly small cable systems used within a single organization. The required equipment is roughly the same as would be required for a cable TV headend, but the scale in terms of number of channels, size of antennas and smaller, and services needed bi-directional communication, such as internet access and pay-per-view are absent.

Cable set-top boxes

The set top box is a device that takes the cable signal in and selects one or two programs of interest. These programs are then passed on to the television, typically using a video cable. Thus, the set top cable box is simply a receiver for the particular transmission frequencies used by cable TV. Since 2007 it has been commonplace for television sets in the United States to include a receptacle for a **CableCARD**, which is a microcomputer in the form of a PCMCIA card that interacts with the cable signal. The CableCARD is thus essentially a set top box built into the TV set, and it also provides a standardized mechanism for encryption and digital rights management (which is discussed at greater length in another chapter). Further details on the way the CableCARD system is used for controlling access to specific programming is discussed further in the section on Conditional Access Control (page 170).

Cable devices are classified as S-cards and M-cards depending on whether the can process a single incoming program stream, or whether they can simultaneously handle multiple input streams. Devices often have to handle both decryption of incoming content as well as encryption of outgoing content to assure copy protection rules are obeyed. As an example, the block diagram for a device that can simultaneously handle six input streams is shown in Fig. 22.

Switched Digital Video (SDV) is a recent technology for compressing more channels into the available bandwidth of a cable stream. It only works with HDTV content, however, and thus is not compatible with older set top boxes or older CableCARDs, even when those older boxes are digital standard definition. That's because it is build around bi-directional data transfer: requests from the CableCARD to the operator determine what streamed data is sent to a viewer, as opposed to just sending all channels all the time and letting the TV pick what it wants. Instead of sending all channels at all types, only a selected subset of the total set of channels needs to be transmitted. This is especially important with HDTV video, since it requires so much bandwidth usage that not all of the desired channels could be accommodated without the use of SDV.

OpenCable receiver standards

A standard for cable television receivers that do not need to reply to the transmitter has been defined, and is called the **OpenCable Unidirectional Receiver** (OCUR) standard. A more recent standard is the **Bidirectional OpenCable Receiver** standard (BOCR) that additionally provides a specification for interactive features, including, for example subscription-on-demand services. This allows for the creation of hardware devices that can work with a range of cable service providers, with primary emphasis in North America. OCUR specifications describe both how the CableCARD can be used, as well as how data can be transferred out of a cable receiver so that DRM and conditional access restrictions still apply to it. Devices that adhere to the OCUR standard include an MPEG encoder that builds a transport stream from incoming content, even if it is received as analog data, in which case Macrovision flags are extracted and maintained. Encryption for OCUR and BOCR devices is based on the AES algorithm. These devices are required have upgradeable firmware and to support text mode services such as closed captioning. The standard also describes the protocols that can be used to interact with the device remotely, so that other components of a media system can change channels, for example.

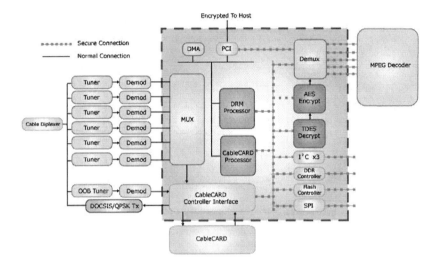

Figure 22: Block diagram of a cable processor which serves as an interface between the cable card and other parts of the system. Figure courtesy of Ceton Corp., applicable to their CTV-9120 M-card processor.

8. Satellite frequencies

Numerous different kinds of signal are transmitted via satellite these days. This includes everything from telephone conversations to seismic data for oil exploration. We will confine our interests here to a few types of transmission: television signals, direct-to-home radio signals, and direct-to-home Internet access. Notably, GPS also uses direct-to-home transmission, but the signals are quite different from the other types since the satellites used are not in geostationary orbits, the GPS broadcast system is administered by the US military, and the data is of a very different nature from what is used for television. Satellite Internet is also a rather different and specialized kind of transmission, having several variations depending how data uploading (from the home to the net) is implemented.

From the point of view of the home viewer, satellite-based television transmission can be classified according to a few major technical factors: the frequency band, the format, and the data encoding. The band refers to the frequency of the radio signals used, but this is almost always also directly related to the power level of the signal being broadcast, since different frequency bands have more-or-less standard transmission powers associated with them. Likewise, there is a close connection between the satellite transmission frequency and the kinds of content it is used for. The key frequency bands will discuss are: the **C-band**, the **K-band** (subdivided into **Ku-band** and **Ka-band**), the **BSS** sub-band and, in passing, the **X-band**.

IEEE Standard Radar Band Designations (see also ITU designations)		
Designation	**Frequency**	**Wavelength**
L Band	1 - 2 GHz	30 cm - 15 cm
S Band	2 - 4 GHz	15 cm - 7.5 cm
C Band	4 - 8 GHz	7.5 cm - 3.75 cm
X Band	8 - 12 GHz	3.75 cm - 2.50 cm
Ku Band	12 - 18 GHz	2.50 cm - 1.67 cm
K Band	18 - 26.5 GHz	1.67 cm - 1.13 cm
Ka Band	26.5 - 40 GHz	1.13 cm - .75 cm
V Band	40 - 75 GHz	7.5 mm - 4.0 mm
W Band	75 - 110 GHz	4.0 mm - 2.7 mm
mm Band	110 - 300 GHz	2.7 mm - 1.0 mm

Names for frequencies

Before getting into the special properties of the different frequencies used for transmission, let's pause for a moment to look at those strange names that are used for them. In fact, all the frequency bands used for satellite reception were once considered for military targeting radar in the first half of the 20th century, before they were used for telecommunications. That's the context in which the names were invented. Sadly, not only are there cryptic names for the different individual frequency bands, but there are also different families of names for the same frequencies, depending on which application community is using the names. The only thing worse than having no standards is having multiple "standards", since that means that in addition to not having consistency you also have different groups that believe they are the only correct ones.

The original nomenclature for radar bands was based on the wavelength of the signals. The longest wavelength signals were called L-band (L for long). These L-band signals had wavelengths of about 23 cm and thus frequencies of about 1.2 to 1.5 GHz. The "other" frequency originally in use had a wavelength that was not as long as those in the L-band, so it was called S-band, for "short" (we should probably be relieved it was not called the NQSL-band, for "No-Quite-So-Long").

The S-band and L-band each have their advantages and disadvantages relative to one another. This led to the development of an intermediate band, which provides a compromise between advantages and disadvantages of the L and S bands, and thus C-band was born (C for compromise). The radar band used to designate targets was called X-band, supposedly for the phrase "X marks the spot".   In Germany an alternative frequency band was put into use, and the wavelengths it used were also regarded as pretty short, and since the word short in German is "kurz", it was and is called K-band (K for kurz).

In practice, frequencies that are just above and just under those that define the K-band wavelength of often used for commercial broadcasting. These are the Ka and Ku bands, where the extra "a" and "u" after the "K" stand for "above" and "under" respectively. There are some other named bands in use too, like S, V and W, but the stories behind them are not so colorful, and they aren't so relevant for home use.   Although we won't discuss them in details, The L ad S bands are used for commercial direct-to-home broadcasting in specific regions. A more complete list of the named frequency bands is shown in the tables below.

Radar Band Nomenclature: International Telecommunications Union (ITU) names	
Band Designation	**Frequency**
VHF	138 - 144 MHz
	216 - 225 MHz
UHF	420 - 450 MHz
	890 - 942 MHz
L	1.215 - 1.400 GHz
S	2.3 - 2.5 GHz
	2.7 - 3.7 GHz
C	5.250 - 5.925 GHz
X	8.500 - 10.680 GHz
Ku	13.4 - 14.0 GHz
	15.7 - 17.7 GHz
K	24.05 - 24.25 GHz
	24.65 - 24.75 GHz
Ka	33.4 - 36.0 GHz
V	59.0 - 64.0 GHz
W	76.0 - 81.0 GHz

Frequencies and Wavelengths associated with different bands (ITU naming scheme)		
Name (acronym)	Frequency	Wavelength (metric units)
Extremely Low Frequency (ELF)	3 - 30 Hz	100,000 km - 10,000 km
Super Low Frequency (SLF)	30 - 300 Hz	10,000 km - 1000 km
Ultra Low Frequency (ULF)	300 - 3000 Hz	1000 km - 100 km
Very Low Frequency (VLF)	3 - 30 kHz	100 km - 10km
Low Frequency (LF)	30 - 300 kHz	10 km - 1 km
Medium Frequency (MF)	300 - 3000 kHz	1 km - 100 m
High Frequency (HF)	3 - 30 MHz	100 m - 10 m
Very High Frequency (VHF)	30 - 300 MHz	10 m - 1 m
Ultra High frequency (UHF) (microwave)	300 - 3000 MHz	1 m - 10 cm
Super High Frequency (SHF) (microwave)	3 - 30 GHz	10 cm - 1 cm
Extremely High Frequency (EHF) (microwave)	30 - 300 GHz	1 cm - 1 mm

Within any frequency band, various different signals can be transmitted, just as many kinds of music can be played by a radio station. The amount of capacity of a frequency band that is used for a transmission is measured by the concept of its **bandwidth**. From a television standpoint, the key division is between signals that are transmitted digitally and those that are transmitted using analog coding. As we have seen, traditional standard definition television broadcasting until the advent of high-definition technologies was based on an analog encoding of signals (although analog signals can also be encoded within a digital transmission).

Finally, in addition to the choice of frequency and the choice between analog and digital signals, it is worth noting that there are various formats available for the actual signal being transmitted. These format choices can include issues such as the choice of encryption method, the way different programs are combined to produce a final signal, or the kinds of audio or video data being sent. Since the encoding format is such a rich and complicated issue, it is discussed all by itself later on.

VHF and UHF frequency bands are used for terrestrial television and radio. SHF and EHF are used for radar, among other things. L-band is used for the Global Positioning System, as well as being used for internal communications in satellite receivers. S-band is also used in wireless data communications (such as 802.11). Let's consider the most commonplace frequency bands for television in greater detail, since they have an impact on system setup.

UHF and VHF

The Very High and Ultra High radio frequency bands have been used for terrestrial television for a long time, and are still used for digital television as well.

C-band

The C-band and the higher K-band frequencies are referred to as radar, or microwave, frequencies, and they are used for satellite television transmissions. C-band transmissions have been in use for commercial television transmission for longer than the other frequencies, and were initially used for internal purposes by broadcasters. For example, major broadcasters could (and still do) send programming from one coast to the other for subsequent re-transmission in public over-the-air programming.

This is called "back-haul" programming. A related kind of programming is called a "wild feed" which is an unscheduled transmission also meant for internal use by networks, typically dealing with ongoing news, sporting events or other spontaneous happenings. This programming includes unedited broadcasts of news and sporting events that regional broadcasters might edit to extract clips and footage for later use. Tapping into such programming was, at one time, very enticing since the "feeds" sometimes included unedited content that was not otherwise accessible, and was free of commercials. Since the technology was initially very costly, and thus not available to many home users, the transmissions were often unencrypted and open to anybody with the means to receive them. As time has passed, the fraction of unencrypted signals in the C-band has diminished progressively, but there are still large numbers of unencrypted channels including wild feeds. Residential C-band use in the Americas is diminishing as Ku-band use increases. In other regions, and Africa in particular, C-band transmission remains a critical part of the telecommunications infrastructure and will probably be used even more in the near future. Among the several reasons for this two are notable. C-band satellite access is less costly that the more popular Ku/BSS band access, making C-band television telecommunications cheaper for a provider to operate. In addition, C-band frequencies are less susceptible to interference from rain (a phenomenon known as "rain attenuation" or more commonly as "rain fade"). Since Africa, in particular, is subject to very heavy downpours, this improved immunity to rain fade is an advantage for C-band.

C-band transmissions are sent at much lower power levels than K-band signals, and as a result require much larger antennas. This is partly because C-band broadcasting is an older technology. A typical C-band dish has a typical diameter of 4 feet (1.2 meters) and can vary from 3 feet to 9 feet (1 to 3 meters), and sometimes even as much as 12 feet or more, depending on available space and the signal strength of the satellites being sought. In general, a 4 1/2 foot (1.2 meter) dish is a requirement for receiving television programming. Since this is pretty large, C-band antennas are often referred to informally as BUD antennas (big ugly dish antennas). With an antenna this large, several problems can arise: the physical weight strain on a supporting building, casting a large shadow and blocking sunlight, being aesthetically unpleasant, or catching wind (thus increasing support problems). To help relieve these problems, many C-band dishes are made of wire mesh instead of solid metal. This works due to the way microwave signals (and electromagnetic radiation in general) interacts with an antenna. So long as the holes in the mesh are much smaller than the wavelength, the signal will be "caught" (just like catching fish with a net). The wavelength of a C-band signal is about 2.2 inches (56 mm) and thus a mesh spacing half this size can still efficiently catch the signal while having much less weight than a solid antenna and letting wind and letting sunlight pass through.

Ku band

Ku-band signals are microwaves in the frequency range of 12 to 18 GHz. As discussed above, the term Ku refers to the "Kurtz-under" band. The Ku and Ka bands are used in preference to the K-band itself, since microwaves in the K-band is actively absorbed by water molecules. This makes the K-band just right for use in microwave ovens, but not so great for satellite transmission. Even the Ku-band can be blocked by heavy rain (about 100 mm/hr) or snow, and this phenomenon is known as **rain fade**. Rain fade is fairly uncommon in North America, if the satellite is clearly in view and the satellite dish has been aimed correctly. Statistical analysis of rainfall patterns and signal strengths suggest that fewer than one per cent of rain storms should cause rain fade if the system is installed properly. Rain fade is more likely when one is receiving transmissions from satellites that are barely visible (i.e. not "overhead"). This is both because the signal may be weaker and also because the signal path through the atmosphere is longer and thus the signal has a better change of being absorbed somewhere along the way from the satellite to the antenna. Rain fade is a more serious problem in some parts of the world where very heavy rain is commonplace.

The Ku-band is used very extensively for satellite data in Europe and the Americas and has, to a large extent, displaced the use of the more cumbersome C-band. It is used for commercial programming intended specifically for home users, but also for business-to-business transmissions including network backhauls (raw unedited footage) and for NASA's tracking data relay. A particular sub-range of the Ku-band is called the Broadcasting Satellite Service (BSS) band, and is discussed below.

Based on the strength of their transmissions (about 20 to 120 watts per transponder), the satellites transmitting in the Ku band have to be separated from one another so that their transmissions do not interfere with each other. The geostationary band over North America, Europe and Asia is pretty much completely filled. There are about 22 Ku-band satellites that have coverage of all of North America.

BSS broadcasting

The Broadcasting Satellite Service band (BSS-band) band is made up of frequencies from 12.2 to 12.7 GHz, and refers to a sub-range of the Ku-band used for direct-to-home commercial broadcasting. Broadcasts in the BSS band are typically transmitted with high-power satellites (up to 240 watts per transponder) allowing for the use of relatively small dishes. While the Ku-band, in general, is subject to interference from heavy rain, the BSS-band is sometimes regarded as being more resistant to rain fade; this is a result of the high power transmission power not the properties BSS sub-band itself. As a result of the power level, the satellites using the BSS-band are spaced about 9 degrees apart from one another. At present, essentially all the orbital slots for BSS-band transmission over continental landmasses are filled up.

Many large companies (such as the television providers DirecTV, Dish Network, Bell ExpressVu and Astra) that broadcast high-powered encrypted signals directly to home users use the BSS band for digital direct-to-home broadcasting, and thus the BSS band is sometimes inaccurately referred to as the **DBS-band** (which properly stands for **direct broadcast satellite**). This tie between the BSS-frequency band and DBS usage is especially inappropriate since some DBS broadcasters (such as BSkyB Ltd), especially outside North America, use different parts of the frequency spectrum.

Since the BSS-band is just a set of radio frequencies, they can be used for either analog or digital transmissions. In practice, the television signals transmitted on the BSS-band are almost exclusively digitally encoded.

Ka-band

The Ka band spans frequencies in the range from about 18 to 40 GHz, right next door to the Ku band, but only the portion from 20 to 30 GHz is used for telecommunications. The Ka band has only recently been tapped into for residential television broadcasting and only a few satellites are currently broadcasting accessible signals in the Ku band. In terms of the types of dish and properties of the signal, the Ka band is essentially the same as the Ku-band, and there are hybrid dishes that receive both Ku and Ka band signals simultaneously.

Like the X and K bands, Ka band is also used in portable radar systems used to measure the speed of cars. Ka band has become increasingly popular for police radar, since it appears to be more immune to noise than X or K band signals. It is typically used in portable detectors to measure the speed of a vehicle when it is 1 to 5 miles away. Radar jamming devices in the Ka-band are also commercially available.

X-band

The X-band was extensively used for automotive radar detectors to measure the speed of passing vehicles. Its use for this purpose in the United States, however, has largely been replaced by K-band detectors, and more recently by Ka-band units. A portion of the X-band is also used for terrestrial broadcast in some regions, including radio broadcasting in some parts of North America.

The X-band is also used for communication, in particular military communications and fire control (controlling the firing or launch of military weapons). Due to its short wavelength, it is sensitive to fairly small features on an object of interest, and thus can be used to discriminate between different objects based on the way they reflect or absorb signals. The US Pentagon's existing X-band platform, known as the Defense Satellite Communications System, (DSCS), is aging and no longer capable of meeting the military's needs, according to representatives of the US Defense

Information Systems Agency (DISA). Its replacement is under development and is called the Wideband Gapfiller.

9. Digital Data Encoding: DVB

The average home television viewer does not need to understand the details of how television signals are encoded and combined. This section is for those who really wish to understand the internals of digital satellite television. This information is particularly useful for those who may which to construct or modify software that interacts with digital video data.

Essentially all video broadcasting is based on sending and combining together several different kinds of signal at once. These signals can include video, still pictures, text and other data. While this is especially true of digitally encoded satellite broadcasts, it even applies to older standard analog television transmissions as well. The combination of audio and video in traditional analog television broadcasts is an example of sending two different signals at the same time.

Quick Summary:
The DVB protocol specifies how all the digital packets are arranged. They can be seen a a complex set of linked lists.

DVB, MPEG and ATSC are all closely related at this level.

Even the traditional NTSC video signal alone is a combination of two different kinds of image information combined together. As discussed earlier (on page 15), the original transmission system for NTSC analog video used only black and white TV signals (which was all that could be displayed). When color display technology was invented, this extra information had to be included in a manner compatible with existing television sets. It was (and continues to be) transmitted as an additional optional message mixed into the original brightness-only transmission. This mixing of signals, more technically known as **multiplexing**, is analogous to the packet interleaving of digital video using **MPEG-2** that we will discuss below.

Packet streams

With the advent of digital broadcasting, the practice of combining several messages (i.e. kinds of content) into a single compound signal became much more extensive. This is partly because digital data transmission often breaks a signal into many small distinct data packets for reasons of reliability. Each of these packets contains information describing its individual content. Once a signal is in this form, it is very easy to insert different kinds of additional packets into the sequence of items already being transmitted, and the receiver can just pull out the ones is needs and ignore the others. In fact, most data sent even on a single frequency contains a group of different programs combined in this way.

Combining programming in a transmission can be outlined as follows. Any single program is chopped up into successive packets (like the chapters in a DVD movie, but far smaller), and then packets from the different programs being combined are transmitted one after the other. Each of the programs going into the combining is a separate **program stream**. To allow the programs to be found, identified and re-assembled, various extra information in the form of further additional program streams is also sent. This is used to decide which of the chopped-up components go together. In this way a composite TV broadcast can be split into separate steams representing video, audio in different languages, as well as text and explanatory information, and then these can later be reassembled.

Here's an analogy from (almost) everyday life. Recall that early in the 20th century, you could find serialized stories by Charles Dickens and other writers in the newspaper. Every day another page of the story would appear. If you bought the successive issues of the newspaper, and read them in sequence, you could read the whole classic story of <u>A Christmas Carol</u> in a form that we can regard today as a "streaming novel" (like streaming video on the Internet today). Now, let's imagine a newspaper delivery boy who is lucky enough to deliver a selection of different newspapers on the same route. The newspaper boy is like a broadcaster that delivers different kinds of programming. Each newspaper is like a different frequency (or transponder) of a transmitter, and you can subscribe to any of the papers, and even switch between them, but you can only read one at a time. Now, a single days' edition of all the newspapers is a bit like an entire satellite, that is sum of all the data it transmits. There is a table of contents up at the front, that tells you what page numbers different stories are on and, if the paper were full of serialized stories, it would tell you what pages you could find the parts of different stories on (such as <u>A Christmas Carol</u> is on page 3). It might even say that along with the serialized story of <u>A Christmas Carol</u>, there are pictures of the characters to be found "in the color insert on page C6" (let's just neglect the fact that they didn't have color inserts back then).

Just the way the newspaper is divided into pages, and a table of contents says where the different items are within the papers, a digital television transmission is divided into packets and a table of contents says what programs are made up using what packets. Unlike the newspaper which is delivered once a day, with just one table of contents per issue, television data pours in continuously and so the table of contents must be re-transmitted again and again every couple of seconds. The serialized stories and corresponding pictures in the color insert in this analogy correspond to what we will call program streams, or more technically **Packetized Elementary Streams** (PES), the pages of the paper to program identifiers (PID's) and the table of contents to the **Program Map Table** (PMT). We'll see all this, and lots more related jargon, in a moment.

The main standard for digital television transmission is called DVB and it is accepted and standardized worldwide. In the United States, Canada, Mexico, South Korea, and Taiwan, however, an alternative scheme has been accepted for terrestrial television broadcasting called ATSC. ASTC is used in television transmission as an alternative to DBV, but fortunately for the sake of developers, both are based on the MPEG-2 (and related MPEG-4) protocol and both are very similar to one another. DVB and ATSC standards define how to combine multiple programs in one transmission, how to attach a table of contents (the Program Map Table), and how to encrypt the data if desired. To further complicate matters, there are also a few other proprietary digital TV transmission and encryption schemes in substantial use, notably **Digicipher-I** and **Digicipher-II** (from General Instrument Company, now owned by Motorola) and **DSS** (from DirecTV, owned by News Corp). These are discussed later in the context of content protection (see page 152).

Since DVB is by far the most prevalent and well-documented worldwide standard, and since the basic ideas are common between most systems, we will consider it for now and will discuss the other schemes only briefly. We will also look at the ATSC standard in some detail, but will largely omit detailed discussion of the proprietary systems due the fact that they are less standardized (and thus could change without notice), a lack of public documentation, and possible legal restrictions on what can be divulged.

In normal television we think of information coming to us on different channels. The notion of a channel abstracts the combination of audio data, video data and other information such as a channel name and close captions. In the context of MPEG data, this is referred to as a program, since for stored data the word channels would not really be appropriate. In the case of the ATSC protocol, the term that is used is a **virtual channel**.

DVB stands simply for **Digital Video Broadcast** and it is an open standard regulated by the **DBV Project**, an industry consortium made up of hundreds of companies. It is a protocol for transmitting digital video, audio and textual data over noisy multiplexed links (multiplexing means sending several programs at once). The DVB Project claims there are over 200 million DVB receivers deployed around the world today. Since the DVB protocol is specifically designed for transmission media that may have packet losses (i.e. noise due to interference), the protocol includes provisions for how to cope with missing data. There are also many sub-standards within the broad DVB umbrella that deal with subtopics including the use of specific media, such as satellite transmission, versus over-the-air broadcasting. In earlier revisions of the DVB protocol, all digital video sent in the DVB signal was encoded using the MPEG-2 video standard, a video encoding and compression system used for many different purposes and media. DVB also includes provisions for little-used MPEG-1 layer I audio transmission, Dolby **AC-3** formatted audio, and **DTS**-formatted audio. More recently, the standard has been augmented to support the transmission of data in the more recent MPEG-4 video standard as well. Just to get the confusing mixture of terms fully laid out, let's note that the acronym MPEG stands for Moving Picture Experts Group, and the standard is now maintained by the ISO (the main international standards organization) and thus MPEG-2 could equally well be referred to using its less popular official international designation **ISO 13818-1**. Phew! As an aside, there is also a less widely used MPEG-1 standard that predates MPEG-2. Likewise, MPEG-4 is an even more recent video standard, and among other things it can allow even more compression than MPEG-2, and as a result is being used for many newer applications particularly including portable video players and mobile devices like video-enabled phones. A hybrid (not quite standard) scheme called MPEG 1.5 also exists for specialized applications such as CNN's TV broadcasts that are provided specifically for use in airport departure lounges.

Streams

The Packetized Elementary Stream (PES) is the basic building block of MPEG-2 (and thus DVT or ATSC) data transmission. An **Elementary Stream** (ES) is the raw data that constitutes one kind of information, such as an audio track or a video signal. Once this is chopped up into a string of individual data packets, it becomes a **Packetized Elementary Stream**.

Now, just when you think you're done with terminology, we need to also note that MPEG-2 and MPEG-4 are very general standards that not only have lots of options, but also have a couple of important variations; consider them to be like spoken languages with different dialects, if you like. These variations are set out in "part 1" of the official standard, which deals with multiplexing (combining) and synchronizing data. To keep the complexity manageable, let's consider just the MPEG-2 standard (not MPEG-4, which is quite similar at this level of detail).

There are two kinds of MPEG-2 data stream that we need to be concerned with. These are the MPEG-2 **Transport Stream** (TS) and the MPEG-2 **Program Stream** (PS).   Each one is a technique for taking multiple Packetized Elementary Streams and combining them together, so that a packet from one PES is sent, possibly followed by a packet from a different PES. A system like this that sends one kind of data, followed by another, is referred to as **time division multiplexing**. It divides the available time using the transmission system between the data for the different Packetized Elementary Streams. As the packets are combined, each one is preceded by a header that describes which Packetized Elementary Stream it came from, so that the individual packets can be sorted from one another and the separate streams can be reassembled later.

The Transport Stream (TS) variant is specifically for less reliable communication media where some of the data may get lost in transmission (for example like a satellite transmission where poor reception may cause packet losses). In a Transport Stream, packet has about the same size, and error correction is an important consideration. One advantage of using small packets is that if a few of the packets are unrecoverably damaged, and thus lost from the stream, only a small hunk of data is lost.

The Program Stream (PS) variant is intended for more reliable media, like a DVD, where things should always work fairly reliably, and we do not need to tolerate so much data loss. As a result, the size of the individual data packets can be much larger than with a Transport Stream. If one gets destroyed, it may have a noticeable disconcerting effect for the viewer, but we don't expect this to happen. The advantage of the Program Stream format is that it uses the transmission media more efficiently since less overhead is needed for so many of the little packet headers and ancillary data. As a mnemonic trick we can note that the phrase "Transport Stream" places emphasis on the (unreliable) way the data is *transported*, while the "Program Stream" places more emphasis on the *program* itself and thus doesn't worry as much about the medium and its faults.

DVB protocol variations

The DVB protocol comprises three main standards, each applying as a function of how the digital video is being broadcast and used. These three variations are **DVB-S**, used for satellite broadcasting, **DVB-C** used for cable broadcasting, and **DVB-T** used for over the air broadcasting (the "T" is for terrestrial which makes it sound like the signal is being sent through the earth, but terrestrial is the normal term used in contrast to satellite broadcasting). These three systems have many common features but also some significant differences, largely based on assumptions regarding how much signal loss is likely as well as how much bandwidth is available.

Let's focus on the DVB-S signal used with satellite transmissions, and we can consider the other alternatives later on. Since the signal is digital it can naturally be organized into bits, bytes and larger groups of bits called packets, which can be large or small. Each individual packet contains a single kind of data, and to send multiple types of data in a DVB transmission the different data types are mixed together by sending different kinds of packets one after the other, but never by mixing data within a packet. Anything bigger than a packet, like a picture, is divided into a series of these packets. Anything smaller is put inside a single packet, and the extra space is leftover (which is not much of a problem since the packets are fairly small). The process of sending more than one transmission at once, first a packet of program one, then a packet of program two, than another of program one, is a form of multiplexing, more specifically time division multiplexing since we are sending different things are different times. When time division multiplexing is used to transmit satellite data, it's called a **Multiple Channel Per Carrier** broadcast, or **MCPC** for short. When broadcast companies are sending data between their own private facilities, for example between different production studios, they sometimes forgo the time division multiplexing, just blast down a single program on each slightly different radio frequency, and that's called **Single Channel Per Carrier** (**SCPC**) broadcasting. The NBC broadcasting network in the USA, for example, uses this kind of SCPC broadcast to send MPEG-2 encoded programs between its different facilities on the GE-1 Ku band satellite. These "back haul" programs are subsequently sent to consumers as part of normal DVB broadcasts.

Addition members of the DVB class of protocols are **DVB-H**, DVB-**S2** (also written as DVBS-2), and ATSC. DVB-SH is a variation of the DVB-T terrestrial broadcast protocol specifically aimed as handheld mobile devices like cell phones allowing for the combination of television and internet data streams. Closely related variants also for transmitting data to handheld devices are DVB-H (which is slightly older) and DVH-H2. These protocols for handheld devices will probably gain substantial ground in the near future, and use a combination of broadcast signals to attempt to achieve continuous reliable service.

DVB-S2 refers to "second generation satellite transmission" and it is an extension and update of the DVB-S protocol to optimize it for better performance, high definition television (HDTV) support, and a few extra services. The most important differences between DVB-S2 and DVB-S relate to support for adaptive changes in transmission parameters, which allows DVB-S2 to achieve higher performance than DVB-S while being based on the same underlying MPEG-2 (or MPEG-4) encoding mechanisms. DVB-S2 also supports additional transmission schemes beyond just QPSK (discussed below), and more robust error correction. DVB-S2 is not intended as a replacement for DVB-S for the foreseeable future, but as a way of supporting additional applications. That said, several broadcasters of satellite television, particularly in Europe, have already adopted it. DVB-S2 provides higher quality signals and uses H.264 encoding.

ATSC is a relative of, and alternative to, DVB-T that was created for the North American market. There are also a slew of other specialized DVB protocol variations and specialized standards related to signal security (**DVB-CPCM**), networking (**DVB-IP**), interactive-TV for the "multimedia home" (**DVB-MHP**) and other specific issues. These additional extensions are, by and large, supplements to the basic DVB protocols. As an example of one of these supplements, DVB-MHP relates to the API and required functions for a Java-based software suite that would support DVB data and deliver video recording functionality and other features as part of a generic multi-media platform for home use.

One final variation of the DVB standards collection is **DVB-ASI**, where ASI stands for **asynchronous serial interface**. This is a variation of the

DVB protocol used for video encoding and editing without concern for long-distance transmission (i.e. it is for video transmitted from one box to another by coaxial cables, typically on a single desktop or rack). It's used as a generic format to specify what video processing equipment can handle, and it competes with other (older) video encoding standards like the **SMPTE** serial digital interface (which comes in variants SMPTE **259M** or **292M**). SMPTE 259M is a system for encoding and transmitting uncompressed standard definition digital video and supplementary data (like captioning) to provide synchronization and error correction. The SMPTE 292M standard is similar, but is used for HD video. DVB-ASI uses pretty much the same architecture and data packets as DVB-S, allowing for the transmission of compressed video, audio and extra data, but specifies how the data is carried on the cable (based 8B/10B amplitude modulation encoding using 8 different pulse amplitudes), and that the data rate should be (270 MBps). The 8B/10B system is a widely used encoding standard that uses 10 data bits to encode every 8 signal bits. It is an error correcting code with in interesting property that it also allows for "DC balance". That means that over a series of transmitted bits, it is possible to assure that the total number of zeros and ones used is equal, and thus if zeros and ones are coded as offsets from a ground voltage, the signal has essentially no net (long term) average voltage offset.

When we consider DVB data transmission systems we can break the technology into three main kinds of problem: the hardware, the low-level (analog and bit-level) transmission protocols implemented by the hardware, and the packet-level digital data protocols carried on the radio signal. Right now we will focus on the packet-level digital data, but lets say a few words about the radio communications first. The DVB-S transmissions are sent using a modulation scheme known as **QPSK** (Quaternary Phase Shift Keying). This scheme is especially well suited to signals with high noise levels. The amount of noise or interference in a signal is measured by the **signal-to-noise ratio** (SNR), where smaller numbers are bad. QPSK attempts to provide some immunity to noise. It sends multiple bits per tick of the signal clock by using the signal **phase** as an encoding channel. In addition, the data encoding at the lowest levels includes error correction mechanisms that are able to repair either bit errors (within a byte) or byte errors. The particular mechanisms use a trick called "convolution interleaving" to distribute bits to avoid a noise burst and include punctuated convolution coding and Reed-Solomon block coding, both of which are classic high-performance error correction methods used in CD's, hard disk drives and DSL transmission as well. The basis of all such error-correcting codes is the transmission of extra bits that allow one to determine if a signal has been damaged in transit, and then which can potentially be used to repair the damage. The number of damaged bits can usually be measured, and the number of errors is reflected in a figure called a **Bit Error Rate** (BER) that is reported by some receivers.

DVB-S2 transmission used for some HDTV broadcasts uses even more bandwidth than can be normally achieved with QPSK. For this reason, DVB-S2 systems sometimes use **8PSK** for HDTV broadcasting (where QPSK encoding could also be called 4PSK, since the Q stands for quaternary -- having four). 8PSK modulation uses the same principles as QPSK, but it can achieve somewhat higher data rates, although this comes at the expense of somewhat less noise immunity.

MPEG-TS data streams

As we noted above, the data sent in a DVB transmission is made of a combination of individual broadcasts, or elementary streams, encoded using the MPEG-2 TS protocol. The whole selection of streams and all the data to sort them out is called a **bouquet**. Each of the streams of data that make up a program is sent as a sequence of individual packets, which are selected and extracted to make up a particular show you might want to watch. This extraction process leads to a sequence of packets that all belong to one program. The sequence of packets making up the video image data for a single program is called a Packetized Elementary Stream (PES). When the video and audio PES streams are interleaved, this becomes a Program Stream (PS). This is very much like an MPEG-2 file on your computer, but it has been subdivided into separate packets. Also, since this is the Transport Stream variation of MPEG, which is meant to be robust in the face of transmission errors, each packet includes a header with information, including extra sequencing and timing data to help deal with errors.

The packet header

Each packet of the MPEG-TS stream is 188 bytes in length. This is expanded by an extra 20 error correction bytes to a total of 208 bytes for the ATSC variant of MPEG-2. A single packet consists of a header that describes what kind of content in contained in the packet, and a payload that may be, for example, audio or video data.

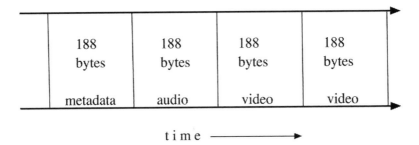

Figure 23: MPEG-TS packet stream.

The packet header includes the following key components:

o A single byte with fixed hexadecimal value 0x47 to denote the start of a packet (known as the **sync** byte) .

o A 13-bit program identification (PID) number. This a label on the packet saying what kind of data it holds, and perhaps what content it belongs with.

o A 2-bit value, the scrambling control bits, indicating if the program has been encrypted (known as scrambled). Encrypted data are used by many commercial program providers, so that viewers need to purchase a subscription to decrypt and watch the programming.

o An optional 42-bit synchronization counter that helps compensate for missing data.

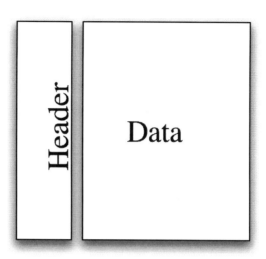

Figure 24: Division of a packet into a header and a data block. The header specifies the structure of the data block, but not fully since some groups of different data blocks share a common header.

Several additional data fields are also present in the packet header.

The scrambling control bits describe the encryption status of the packet. Only some packets types, such as audio and video can be encrypted. For some broadcasters, even though the program itself may be encrypted, various kinds of metadata such as the program guide may not be. If a packet is encrypted, additional information regarding the encryption system must be recovered from packets whose content specifically relates to conditional access. The scrambling control bits can be interpreted as follows: have the following values and associated meanings:

Scrambling control bit pattern	Meaning
00	Unscrambled content
01	Not scrambled, but state may have additional significance specific to the broadcaster.
10	Packet payload data is scrambled with "even" key.
01	Packet payload data is scrambled with "odd" key.

Classes of data packet

In addition to the program itself that is being transmitted, the data stream includes several types of non-program data (sometimes called meta-data) that describe and supplement the program content. This data comes in the form of "service information" (SI) tables that index and tie together the data and its properties. These various tables are groups into two classes: the "PSI" tables that are part of the basic MPEG-2 specification and which could be part almost of any MPEG stream, and the "SI" files that are specific to DVB broadcasts in particular (i.e. which can be seen as extensions to MPEG). Collectively, in the context of DVB video transmission, these PSI and SI tables are often refereed to together as simply SI tables.

The ATSC variant of DVB, used for cable TV and terrestrial broadcasts in North America, uses a set of tables called PSIP tables instead of SI and PSI tables, but the PSIP carries essentially the same information. PSIP tables, however, store the information using a slightly smaller number of different program ID types than DVB (due to technical restrictions in the hardware specification). As a result of using a smaller number of PID streams, the ATSC protocol is forced to combine more than one type of table under the same PID. This can make the logic to identify data tables slightly more complicated. These tables are used to allow an over-the-air tuner to find stations, to provide a information for a program guide, to assign names to channels, and to specify and control data encryption.

The collection of packets in the stream can be viewed as an interlinked data structure making up a simple tree. Each type of packet has a PID value that identifies the kind of information it carries. The root of the tree is the critical structure called the **program association table** (PAT) and it is always identified using a fixed PID value. It is the most important part of the set of service information tables, since it is the key to sorting everything else out. It provides the program ids (PIDs) of other kinds of tables, and these in turn can provide information on still others.

Most software that handles DVB or ATSC data has a low level component that scans the transport stream and selects packets with PID numbers of interest. This, in essence, reconstitutes the elementary streams that are assembled to make the transport stream. The DVB driver in the Linux operating system, for example, contains a mechanism in the operating system DVB driver that allows an application to register to receive packets with particular PID's of interest.

Specific data tables

The different PIDs and the tables they identify have specific functions and store different kinds of data. These are described below.

program association table

The program association table (**PAT**) is the key to sorting out the packet types that make up a stream. It always has the special program id number (PID) of zero. It contains a list of all of the program map table id numbers (PIDs) for the different programs that make up the transport stream. Since it is so important, the DVB-S specification requires it to be retransmitted in the transport stream at least once every 100 milliseconds (1/10 second).

The detailed makeup of the table is illustrated below. Although the precise details are very technical, a few key features are worth noting. The **table id** provides a supplementary ID number that identifies a sub-type of data (for different types of data within the table that share a common PID). The **transport stream id** is a number that uniquely identifies this transport stream relative to all others that share the same network ID.

```
program_association_table {
    table_id                        8 bits
    section_syntax_indicator        1 bits
    '0'                             1 bits
    reserved                        2 bits
    section_length                  12 bits
    transport_stream_id             16 bits
    reserved                        2 bits
    version_number                  5 bits
    current_next_indicator          1 bit
    section_number                  8 bits
    last_section_number             8 bits
for (j=0;j< N;j++) {
        program_number              16 bits
        reserved                    3 bits
if (program_number=='0' ) {
        network_PID                 13 bits
        }
else {
        program_map_PID             13 bits
        }
    }
    CRC_32                          32 bits
}
```
Data structure layout of program association table (PAT)

program map table

The data packets that make up the program map table (**PMT**) provide information about a specific program (also called the Program Management

Table in some literature). In particular, it provides information regarding the format of the transmission and, most importantly, the program id number (PID) of the audio and video packets that make up a program, PIDs of packets providing supplementary program data. Simply put, the PMT provides the key to finding all the other streams, and hence data packets, that carry the actual program. As an illustration of how these data tables are constructed, the following C-like data declaration describes the detailed structure of the PMT.

```
program_map_structure {
    table_id                 8 bits, always 2;
section_syntax_indicator
    1                        1 bit, always 1;
    0                        1 bit, always 0;
    reserved                 2 bits;
    section_length           12 bits (number of bytes
                                 in the
                                 rest of this record);
    program_number           16 bits (program this PMT
                                 applied to);
    reserved                 2 bits;
    version_number           5 bits;
    current_next_indicator   1 bit (if 0, this map
                                 not valid yet);
    section_number           8 bits, always zero;
    last_section             8 bits always zero;
    reserved                 3 bits;
    PCR_PID                  13 bits (PID value of
                                 associated PCR);
    reserved                 4 bits;
    program_info_len         12 bits (number of bytes
                                 of following
                                 descriptors);
for (i = 0; i < N2; i++) {
    descriptor() [descriptors that apply to the
    whole program, see below];
    }
```

```
for (i = 0; i < N1; i++) {
        stream_type              8 bits (type of PID);
        reserved                 3 bits;
        elementary_PID           13 bits (PID number);
        reserved                 4 bits;
        ES_info_length           12 bits
                                 (number of bytes of
                                 following descriptors);
for (i = 0; i < N2; i++) {
    descriptor()
    [descriptors that apply
      to the current elementary
      stream see below];
        }
    }
    CRC_32                       32 bits;
}
```

Data structure layout of program map table (PMT).

The contained descriptors are records that can describe a large variety of different item types, about 35 of which are standardized and some 200 or so are broadcaster-specific. They include things like the related video streams (via the video stream descriptor), the conditional access system (encryption), and the copyright status of the program. The video descriptor, for example, indicates if the video is MPEG-1 encoded, if the frame rate is constant, or if the video is composed only of still pictures. Like the PAT, the DVB specification requires it to be retransmitted in the stream at least once every 100 milliseconds (1/10 second)

Network Information Table

The Network Information Table (**NIT**) provides tuning information on various different transport streams. It allows a receiver to find out what other streams can be received on different frequencies by a suitable receiver. This, in turn, allows a receiver to find additional programs once it has an initial one. The NIT can be used by generic receivers to scan for multiple channels and find the various streams (on different transponders) that the receiver can tune to. It is transmitted in the stream at least once every 10 seconds.

Conditional Access Table

Packets that make up the Conditional Access Table (**CA** or CAT) are required for scrambled (encrypted) data streams. This table is typically used to tie the broadcast and its encryption to one or more Entitlement Management Messages streams (EMM messages) that may be used to control access. This typically includes a PID used to identify the packets making up a stream containing the Entitlement Management Messages.

A CA descriptor can also appear within the PMT described above. In this case the CA information provides information on Entitlement Control Message information (ECMs) that is specific to a single elementary stream.

The CAT can include some number of bytes whose meaning and internal structure is free to be defined by each data provider and is not specified in a standard way.

Beyond PSI: SI tables

The PAT, the PMT, the NIT and the CA tables make up the MPEG-specified Program Specific Information tables (the PSI). These alone let you get the program you want to watch out of the transport stream. The DVB specification provides extra specifications and describes a couple of extra tables, as follow.

Bouquet Association Table

The Bouquet Association Table (**BAT**) groups together different kinds of data and PID's that may span different transport streams (and hence different transponders) and relates them to one another. It can even (hypothetically) relate programming and data on a satellite broadcast to programs delivered in some other way, for example by cable TV. This table allows for a consistent presentation of material (such as a guide) regardless of technical details such as the fact they are in different streams.

Service Description Table

The Service Description Table (**SDT**) provides descriptive information such as the names, the countries of availability, and the languages used for the programs in the current and/or other transport streams. In can include subtables and it is always on PID 0x11 (hexadecimal 11). The function and contents of the Service Description Table and the Bouquet Association Table have some overlap, but the SDT is a required table (unlike the BAT). The SDT table includes a flag to identify free-to-air (unencrypted, unprotected) content that can be displayed on any receiver.

Event Information Table

The Event Information Table (**EIT**) is used to provide information on scheduled programming being sent in various streams. This includes information on television programs such as when they start and end, whether they are currently running, or starting soon. In some countries this table is mandated to be unscrambled while in some regions, and for some providers, the information is scrambled. In fact, the EIT is the only one of these tables whose content can be scrambled, according to the DVB specification. It is not clear what benefit is derived from scrambling the guide, which only helps to advertise available services.

The event information for the current stream is retransmitted at least once every 2 seconds. The longer-term event information, including events on other channels, is broadcast event 10 seconds.

Time and Date Table

The Time and Date Table (**TDT**) provides information about the current data and time specific in UTC (Universal Time Code, formerly known as

Greenwich Mean Time—GMT). This can be used to set the clock on a receiver or synchronize clocks. The information is encoded in a format called MJD (Modified Julian Date) which was developed by space scientists into the 1950s and is generally used for telecommunications. The Julian date is defined as an offset in days from noon, January 1st, 4713 BC (i.e. the year -4713) and has been in use since the 16th century. The MJD can be obtained by subtracting 2,400,000.5 from this value, which corresponds to resetting the system to have an initial day (day zero) at noon Nov. 17, 1858. The Time and Date Table is broadcast at least once every 30 seconds.

Time Offset Table

The Time Offset Table (TOT) is used to provide time offsets, which is to say adjustments, between the local time and the time zone used for broadcasts.

Running Status Table

The Running Status Table (RST) is used at the beginning or end of an event (such as a program) to indicate its status has changed. This is meant to allow automated switching to and from programs as they start or stop.

Discontinuity Information Table

The Discontinuity Information Table (DIT) is used to indicate that the stream is a recording (i.e. a playback from disk and not actual streamed data), or for other reason is not made up of normal data running continuously as a regular ongoing stream. This is typically used to make it clear that some tables may be missing or incomplete.

Selection Information Table

The Selection Information Table (SIT) is used to inform a downstream device that Transport Stream information has been removed and the stream may be only a partial stream that is missing some of the usual SI tables.

Stuffing Table

The Stuffing Table (ST) is used to send information that may invalidate (i.e. cancel) information in other tables, for example if a stream changes function over time. Stuffing Table packets can arrive on a number of different PID's normally associated with different kinds of special-purpose tables.

Program Clock Reference

The data packets that make up the program clock reference (PCR) contain absolute timing information on the data in the transmission. They allow a receiver to determine if the packets are being played too quickly or slowly, and if there is missing data (or even surplus data) in the packet stream.   Not that the Program Clock Reference relates to the timing with program, and is not the same as the Time and Date Table packets that allow an absolute system clock to be set.

Program Service Information

Packets the constitute the Program Service Information (PST) can be made up according to the whims of individual commercial providers and can be used to include supplementary data such as electronic program guides (EPGs) that describe the various shows that are on at different times of day.

Standard PIDs for key tables are listed own below.

PID value	Table name
0x0000	PAT
0x0001	CAT
0x0002	TSDT
0x0010	NIT
0x0003 - 0x000F	Reserved
0x0011	SDT
0x0011	BAT (same PID as STD)
0x0012	EIT
0x0013	RST
0x0014	TDT
0x0010 through 0x0014	ST
0x0015	network synchronization
0x0016	RNT
0x0017-0x001b	Undefined, for future use
0x001C	inband signaling
0x001D	measurement
0x001E	DIT
0x001F	SIT
0x1FF7	PAT-E (ATSC only)
0x1FFB	PSIP (ATSC only)

Each table starts with a special ID code for further identification, over and above the information conveyed by the PID. This is critical since some tables share the same PID with other types of tables in the stream. With the ATSC protocol that overloads the PID types even more, this is even more critical. Some of the more fundamental table ID codes are as follows:

Allocation of table id values (DVB and related protocols)

Table ID code	PID it is found in	Name/Description
0x00	0	program association section
0x01	1	conditional access section
0x02		program map section
0x03	2	transport stream description section
0x04 to 0x3F		reserved
0x40	0x10	network information section - actual network
0x41	0x10	network information section - other network
0x42	0x11	service description section - actual transport stream
0x43 to 0x45		reserved for future use
0x46		service description section - other transport stream
0x47 to 0x49		reserved for future use
0x4A	0x11	bouquet association section
0x4B to 0x4D		reserved for future use
0x4E	0x12	event information section - actual transport stream, present/following
0x4F	0x12	event information section - other transport stream, present/following
0x50 to 0x5F	0x12	event information section - actual transport stream, schedule
0x60 to 0x6F	0x12	event information section - other transport stream, schedule
0x70	0x14	time date section
0x71	0x13	running status section
0x72		stuffing section
0x73		time offset section
0x74		application information section (TS 102 812 [17])
0x75		container section (TS 102 323 [15])
0x76		related content section (TS 102 323 [15])
0x77		content identifier section (TS 102 323

Table ID code	PID it is found in	Name/Description
		[15])
0x78		MPE-FEC section (EN 301 192 [4])
0x79		resolution notification section (TS 102 323 [15])
0x79 to 0x7D		reserved for future use
0x7E		discontinuity information section
0x7F		selection information section
0x80		to 0xFE user defined
0xFF		reserved

Digital data using ATSC

In this section, we will concentrate on the ATSC extensions to the MPEG-2 format, and how ATSC differs from the DVB standard. This section, like the previous one, is intended for those who want the details of the technical implementation of video and audio packet encoding. Aside from intellectual interest, this is likely to be of value to those who want to implement or modify software for the reception or processing of ATSC-encoded programs.

The ATSC packet names generally follow the same outlines are those used for DVB or MPEG-2, with the addition of the letter "E" to many of the packet names if they are transmitted via the *Enhanced* 8-VSB radio protocol used for ATSC. For example, the Program Association Tables is known as the **PAT-E** (rather than the PAT used with DVB), and the Program Map Table is known as the **PMT-E**. Several other minor naming variations are also used. The multi-program transport stream used with DVB is referred to in official ATSC terminology as a **Digital Television Standard multiplexed bit stream**.

In addition to data formatting, ATSC places restrictions on the signaling and timing of the data stream. For example, it includes a requirement that successive program map tables must arrive within 400 ms of one another or that the successive PMT-E occurrences should arrive within no more than 1.6 seconds.

The ATSC protocol does not use the Network Information Table, and none is expected to be present.

The Program Association Table PAT-E is identified using the fixed PID value 0x1FF7.   The PID 0x1FFB is reserved for packets carrying system information and program guide information for the full-resolution transport stream, and 0x1FF9 for any reduced resolution versions of the same program.

The most significant difference between ATSC and DVB relates to the system information (SI) tables of DVB, and the corresponding Program and System Information Protocol (PSIP) used by ATSC.

Allocation of ATSC table id values

Table ID code	PID it is found in	Name/Description
0xC0	PMT	Program Information message
0xC1	PMT	Program Name message
0xC5		System Time table (ATSC only)
0xC6		
0xC7	0x1FFB 0x1FFC	Master Guide table (MGT)
0xC8	0x1FFB	Terrestrial Virtual Channel table
0xC9	0x1FFB 0x1FFC	Cable Virtual Channel table
0xCA	RTT	

Other non-DVB transmission systems

In this section, we will briefly discuss a few alternative television transmission systems. In particular, we will discuss proprietary systems owned by one or another single company that are not documented in international standards. These systems are generally not well described in publicly available documents, and are subject to change without notice. As such, the information here should be regarded as subject to change.

DigiCipher2

The **DigiCipher 2** is a program distribution system, but also refers to the encryption and access control system used by that system. Developed by General Instruments and now owned by Motorola Inc. It is used, in particular, on many cable television systems, as well as the Canadian Star Choice satellite broadcaster, and the **4DTV** class of receivers for C-band and Ku-band satellite signals.

The Digicipher 2 (DCII) distribution protocol predates the development of DVB, but shares many common features with it since both are based on MPEG-2 packet delivery. The basic architecture for video and auto is essentially the same for Digicipher 2 as it is for DVB and ATSC, but each has a different approach to the delivery of service information (SI or PSIP packets), which identifies the channels and the relationships between data streams. The set of SI packets used by DCII systems tend to be more tightly inter-related, with a series of tables being needed to receive a channel.

Digicipher 2 transmission is very similar to DVB, being based on QPSK signal modulation. On the other hand, the detailed parameters that define the signal are sufficiently different that the receivers for the two systems are quite incompatible.

DirecTV DSS

The satellite broadcast system referred to as **DSS** is used exclusively for the DirecTV Company's broadcasts. It has in common with other systems the fact that it uses packet of digital packet data, and it uses QPSK modulation, but the specific protocols used to transmit data appear to be very different from DVB and its cousins. The individual packets are 127 bytes in length, which is smaller than those used for DVB. A single table called the Master Program Guide (MPG) is reputed to play the role of the family of SI packets used in DVB transmissions.

At appears that DSS transmissions are gradually being supplanted by DVB-S2 by its broadcaster at they migrate to HDTV.

10. Antennas and Dishes

An antenna collects radio signals from the air and brings them in to the receiver. The antenna is obviously a key element of a receiving system: the performance of everything else depends on it and the quality of the signal it brings in. In the antenna fails to do a god job, it is impossible to make up for it. In this chapter, we consider the antenna requirements and features for satellite television reception.

At a fundamental level, a satellite dish (antenna), a television roof antenna, and the tiny antenna inside a hand-help GPS unit are all the same. They simply pull in high frequency radio waves. The factors that distinguish one antenna from another are the **directional selectivity**, the **frequency selectivity**, the **gain**, the **impedance**, and the physical size and weight. The directional selectivity refers to how much the antenna picks up signals from one direction compared to another (e.g. from the North versus from the East). The frequency selectivity refers to what frequencies it does and doesn't pick up. In informal terms, the gain refers to how sensitive it is in the direction of maximum pickup. The impedance refers to the electrical properties of the antenna, and it's probably the factor one needs to worry about least in practice.

Dish antennas tend to be very good at collecting signals from a specific direction. Thus, they have both very high directional selectivity and very high gain. In fact, the bigger a dish is, the more it will have of each of these two properties. As a result, careful aiming of a dish antenna is more tricky than aiming an antenna for over-the-air reception, and aiming a big dish is more demanding than aiming a small one.

Types of dish antenna and LNB

The natural consequence of all these various features is that not all satellite dishes are equivalent. The antenna you choose, even if it is a satellite dish, select depends critically on the kind of signal you want to receive, your geographical location, and the selection of channels that you are hoping to pull in.

First and foremost, the choice of a satellite dish depends on the part of the broadcast spectrum you are hoping to tune in to. The three common choices you must first select between are C-band, Ku-band and DVB/DBS (digital). As we have seen, C-band is sometimes known as "big ugly dish" and is rarely used for home reception today except in exotic locations or by serious hobbyists. Ku-band is where most of the free-to-air action is: it encompasses analog or digital channels that can be readily and legally viewed without any subscription. Ku-band antennas are almost always quite a bit larger than DVB antennas since Ku band signals are transmitted at lower power levels. DVB signals are the most popular today, and both because they are digitally encoded and broadcast a higher power levels they almost always require a smaller antenna than Ku-band (or C-band, of course). In summary, the sizes of the dish antennas required for C-band, Ku-band and DVB are successively smaller than one another. In any circumstance, you can use a dish that is too large without any problems, but remember that aiming it will be more difficult as the size increases.

The LNB

In addition to the dish itself, the antenna system is composed of 3 other critical components: the **feedhorn** that collects the signal from the dish (much like a funnel), the **low-noise block (LNB)** that selects a subset of the available signals and sends them down the cable, and the **cable** itself. Each of these components also depends, to some extent, on the kind of programming you want to receive and the physical requirements of your setup.

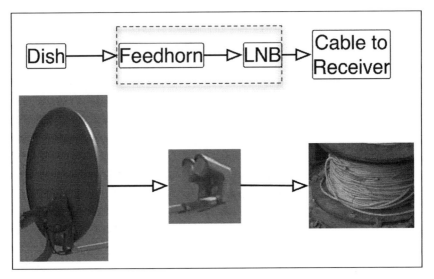

Figure 25: Signal path from dish antenna.

The feedhorn is a cone-shaped component that sits at the end of an arm, at the focus of the satellite dish. The signals that bounce off the dish are aimed into the feedhorn and it collects them at its endpoint. At the endpoint of the feedhorn sits the low noise block, or LNB. The LNB is a combination of a small antenna and an electronic circuit. The circuit runs off about 100 mA of current delivered by the receiver. A satellite dish focuses microwave energy into the feedhorn of the LNB, and an antenna absorbs the microwave signal collected by the feedhorn, thus converting it into an electrical signal. The circuit shifts the frequency of the carrier signal down to a lower one suitable for sending along a cable, and for this reason the LNB is also known as a "block downconverter". This shifting is necessary because the extremely high signal frequencies, in the tens of GigaHertz range, used for satellite transmission cannot be efficiently transmitted along normal cables (without exceptionally large losses). The circuit in the LNB both downshifts the signal to something like 1 MHz by using its own internal timing clock called a local oscillator. The resulting shifted frequency, lower than the one from the satellite itself, but higher than the actual channel you are tuning to, is called the **intermediate frequency (IF)**.

Many receivers need to know the frequency of the IF sent by the LNB when they are configured and installed, while others may ask for the LOF described below. The specific frequencies used for communication between most LNBs and receivers are in the range 950 to 1450 MHz (or 2150 MHz for Ku-band signals, including DVB), which puts them in the L-band. The *difference* between the incoming radio frequency signal and the signal sent along to wire to the receiver is the **local oscillator frequency (LOF)**. The particular formula relating these frequencies is as follows:

IF = absolute_value(LOF - Received Frequency)

for example, if the LOF is 11.47 GHz (11,475 MHz), and the satellite signal is 12.5 GHz, then the IF will be

11475 MHz - 12500 MHz = 1025 MHz.

The typical value for the local oscillator frequency is 10,750 MHz (10.75 GHz) for a digital or Ku-band LNB. For DBS systems it is typically 11,250 MHz. Other common values for Ku-band systems include 12,050 MHz and 10,000 MHz. For C-band LNB systems 5,150MHz is a typical LOF value.

In addition to this frequency shifting, the LNB also amplifies the very weak signal received by the satellite by a factor of about 1000, so that it is strong enough to transmit down the cable. It also shunts the power being transmitted from the receiver into a voltage regulator to make sure it is clean, and then uses the regulated power for the drive circuitry of the LNB.

In many Ku-band, and most DVB systems, the LNB and feedhorn are integrated into a single compact component usually at least 4 inches (10 cm) long and 2 inches (3 cm) wide. Such an integrated unit is properly known as an **LNBF** (with the added F standing for feedhorn), but since that is the only kind of LNB most users ever see (since DVB is the most popular satellite TV format), these are often simply called LNBs also.

The signal quality and frequency stability of an LNB is critical , since the actual signal being amplified is so miniscule. This electrical noise is measured in terms of noise temperature, reflecting the relationship between atomic motion which causes radio noise, as well as heat. The lower the temperature, the better the performance. For many Ku-band LNBs, (which tend to be more noisy than C-band LNB's) this figure is converted to decibels units (dB) instead; a different scale for the same physical measurement. Note that dB is a logarithmic scale, meaning the conversion between noise temperature and dB involves a logarithm function. Using dB also tends to make for generally smaller numbers (making noisy LNB's look better on paper). Some conversion values are listed below.

Noise versus temperature	
Noise factor (dB)	Noise temperature (degrees Kelvin)
0.2	14
0.5	35
0.7	51
1.0	75
2.0	170
3.0	289

If the LNB frequency drifts too much, then the signal can be difficult or impossible for the receiver to extract. As a result, LNBs are sometimes used with integrated phase-locked loops (PLLs) that keep them locked in to a selected frequency. A phase-locked loop is a small electronic circuit used to allow a receiving device to stay locked into a signal once it is initially captured, even if the frequency shifts slightly.   If a signal drifts this is manifested as a phase shift: a difference between the placement of one signal and another, and this phase shift can be used to automatically re-tune, known as staying phase-locked. Typically this can lead to the local oscillator remaining within some 5 kHz of its target frequency, instead of the more common 250 kHz drift one might expected from a good LNB without a phase locking circuit. These "PLL LNB's" can provide a small increment in perceived signal quality, but the difference is probably only worth the increased price in situations where the signal reception is marginal.

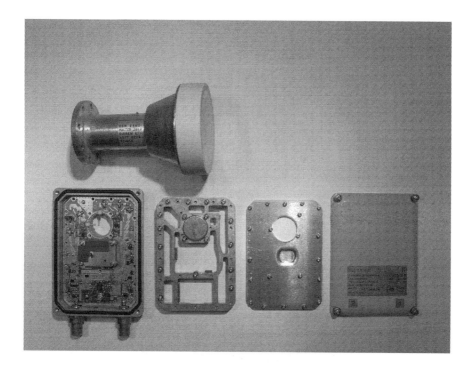

Figure 26: Figure showing a disassembled LNB unit. The amplifier and IF generator circuit can be seen on the lower left

Figure 27: Magnified view of internals on an LNB. The two antenna wires (arrows) for horizontal and vertical polarization can be seen projecting into the circular opening where the feedhorn connects.

Universal LNB

A universal LNB is one that can be used for two very different frequencies, typically for both C-band and Ku-band. These bands require different local oscillator frequencies, and so the universal LNBs have two separate local oscillator circuits. One typically selects between them as if one was selecting a different LNB. This is accomplished by sending a switching signal from the receiver up the coaxial cable to the LNB cable, and is discussed under the section dealing with the DiSEqC protocol (page 236).

Polarization

Electromagnetic radiation, microwaves, TV broadcasts and light are all defined not only by a wavelength, but also a polarization. Think of the waves as being drawn on a sheet of paper; polarization is like the orientation of the sheet of paper.: it can be horizontal (if the sheet lies flat on a table), vertical (if the sheet is held up by an edge), or circular (if the sheet is rotating between horizontal and vertical and back). A more accurate, but complicated, description of polarization would be to consider a vector as a combination of two component vectors, and their phase relationship, but this is technically more involved than what we need to deal with here.

Horizontal and vertical polarizations are both forms of **linear polarization**. In the case of **circular polarization**, then the direction of rotation can be clockwise or counter-clockwise. A signal with a specific signal can be transmitted with one polarization, for example horizontal. A completely different signal can be sent on the same frequency if it uses the complimentary polarization, in this example, vertical.

An antenna needs to be oriented correctly to receive a polarized signal. LNBs can have separate antennas for one or more polarizations inside them, although they either both need to be circular, or both linear. When multiple antennas are used they are connected using different internal circuits and appear to the IRD, for most electrical purposes, like completely different LNBs.

Most DVB signals use circular polarization, although the field is become increasingly diverse. By and large, most Ku band broadcasters use linear polarization.

Multiple LNB's

It has become common for users to want more channels than can be provided on a single satellite. For large dishes it is commonplace to attach a motor to a satellite dish to allow it to be physically aimed at different satellites to receive their broadcasts. For DVB broadcasts, a different approach to receiving multiple satellites has become common. It is based on using a single dish to collect signals from satellites in adjacent slots. One of the disadvantages of smaller dishes (20 inches (51 cm) or less in diameter) is that they are not as selective as large dishes about exactly what they are pointing at. This is why DVB satellites cannot be as closely spaced as those in other bands. A positive side effect of this is that a single dish can pullin signals from more than one satellite at a time. It also makes them easier to aim. Satellite providers have taken advantage of this by developing dishes that can purposely tune in two or more different satellites at once, and focus them at slightly different locations in front of the dish. These are generally used with multiple LNBs and switches that select between them. Modular units with multiple LNBs in one physical unit are the standard solution, but it is even possible to glue LNB's units side-by-side in front of a dish with duct tape if the desired satellites are in adjacent slots.

A typical Ku-band LNB is large enough to that when to are placed beside one anther in front of a 18" dish antenna, their spacing corresponds to the separate between two satellites that are is successive DBS orbital slots (i.e. about 9 degrees apart). In special cases, two LNBs need to be placed closer together than normal, and a monoblock LNB can be used. These are popular particularly in Europe where several commonly-received satellite pairs have a separation of only 6 degrees, for example the popular satellites called Astra 1 and Hotbird are at 13°E and 19°E , or Astra 3A and Eurobird /Astra2. The monoblock is a special combination of LNBs in a single housing, so that their separation is minimized. A typical monoblock is designed with a spacing that corresponds to a six degree orbital separation. This can make alignment of the dish with respect to two different satellites more challenging that usual. This is especially true since even with satellites whose orbital slots differ by six degrees, the distance for an observer may be larger due to the geometric effects of the earth's surface, making the LNB spacing imperfect.

Some LNBs have dual outputs, so that is there are two output cables for the signal, and the LNB can be used to drive two receivers at once. Twin LNBs also have two outputs, but these devices act like two completely separate LNB units, and one can use a switch to select between one output and the other. Some LNBs, such as the Dish Network "Dish Pro" model, have the switching circuitry inside the LNB as well. Switches used to select between different LNB's, either on one dish or with different dishes, are discussed further in the section on *Multiple dishes or* LNBs: DiSEqC (page 236).

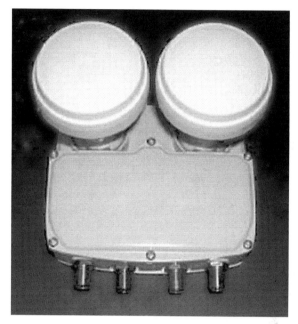

Figure 28: Example of a monoblock LNB.

Dish types

The standard satellite dish has an outer boundary that is parabolic (shaped like a parabola) with an inside that is completely symmetric. The dish acts like a collector or funnel for radio waves, and they are concentrated at the focus of the antenna. This describes a **prime focus** dish, with the focus at the center of the dish itself. For radio waves coming in perfectly parallel into a perfect dish, all the signals would be focused at a single point (this is true for light rays in a telescope as well).

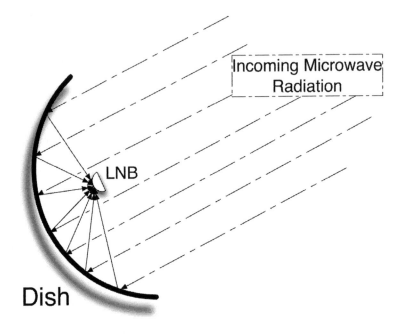

Figure 29: Prime focus parabolic dish with central feedhorn (LNB).

In order to avoid having the pickup device at the focus block the incoming signal, it is usually offset below the dish, as illustrated in the figure below (Fig. 30) and shown in the following photographs.   In practice, the radio signals from different satellites come in from slightly different directions. As a result, as they arrive at a dish they will focused at slightly different points. For this reason there are two ways to tuning in to multiple satellites with a single dish: by moving the dish to point at the alternative satellites, or by placing different receiving LNBs at different points in front of the dish, so that each is that the focus of signals from a different satellite. The second approach (one fixed dish with multiple LNBs) is used extensively for the DVB signals from a single broadcaster (who may send programming on multiple satellites that are in adjacent orbital slots). The approach based on moving the entire dish antenna to aim at different satellites is used mainly with C-band dishes (and occasionally for KU-band).

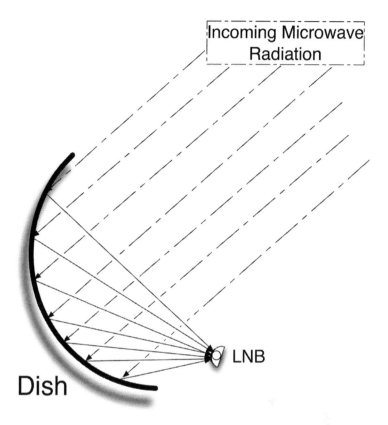

Figure 30: Placement of LNB at offset position.

A more exotic option is to place another reflector at the focus of the dish to better focus the signal into the feedhorn (and LNB). Two such antennas of this type are the Cassegrain antenna and the Gregorian antenna (Fig. 31), both of which use a second reflector in the form of either a hyperbola or an ellipse as the second reflector.

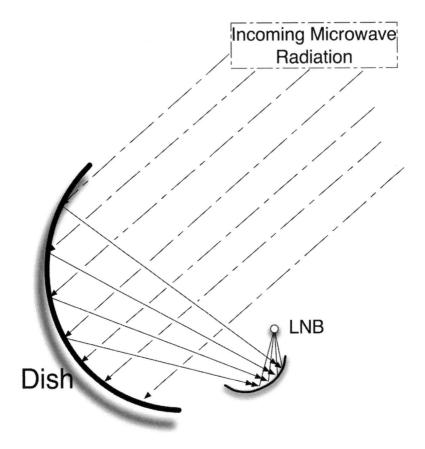

Figure 31: Rough sketch of a Gregorian antenna.

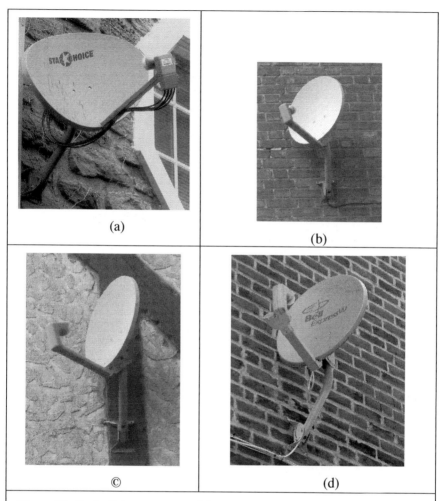

Figure 32: Basic dish antennas used for DVB signals. (a) Parabolic antenna for tuning in multiple satellites at one time. (b, c) Single LNB spherical antenna for a single satellite. Vertical wall mounting. (d) Dual LNB arm with only one LNB used.

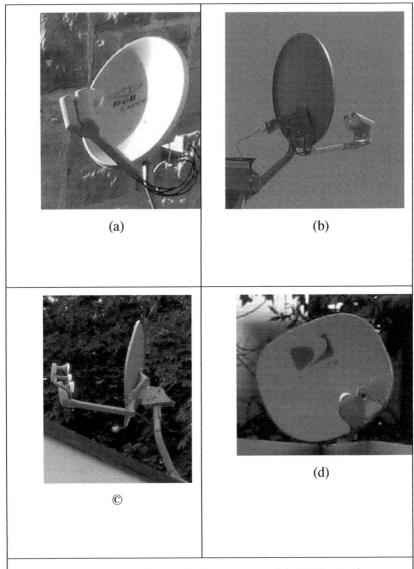

Figure 33: Multiple-LNB dish antennas used for DVB signals.
(a, b) Spherical dish with dual-LNB arrangement. Front and rear views.
(c, d) multiple LNB arrangements for "phase III" dishes used for both Ka and Ku satellites.

As we have already seen, DVB programming is typically received with a "small" dish having a typical diameter of 46 cm (18 inches). The typical size for a Ku-band dish is 76 to 72 cm (30 to 36 inches). C-band dishes tend to be about 2.7 m (9 feet) or more in diameter. For each class of signal, a larger dish will provide better immunity to rain fade, interference, unlucky placement at the boundary of the satellite footprint, and most other sources of signal loss.

The Gregorian antenna is fairly uncommon today, but it can be found in various sizes starting at 55cm and up. The Master Focus brand, for example, includes 55 cm and larger dishes.

Most small dishes are made from solid aluminum, while larger ones are typically made of mesh. Mesh is lighter and doesn't catch the wind, but will still stop microwave signals so long as the holes in the mesh are smaller than half the wavelength.

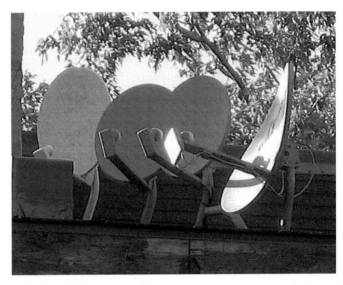

Figure 34: Multiple LNB's and multiple dishes. Sometimes one dish won't be enough.

Types of terrestrial antenna

As we have seen, antennas have particular preferences for one frequency over another, and so there are different types of antenna that are specialized for different frequencies. This is especially important for over-the-air broadcasts, where the range of different frequencies (both VHF and UHF) in use may be quite large. In addition, some antennas are more directionally selective than others. Other than these very significant issues, all antennas work on the same principles and antennas for terrestrial HDTV are the same as those for analog television broadcasting.

Antennas for the reception of terrestrial broadcasts come in several variations. The most critical distinction is between those antennas meant for indoor reception, and those that are meant for use outdoors. Indoor antennas need to be small. Although they come in many different styles, they are all intended for the reception of fairly strong signals. Outdoor antennas, on the other hand, are usually placed on the roof or sometimes on the wall of a building. They are generally much larger and can be arranged in many different configurations both to receive weak signals, as well as to try and reduce (excessive) reception from strong stations (so that weak ones are not overwhelmed). It turns out to be important that all the different channels being received for a given antenna have roughly the same electrical strength once the signal is collected, much like a crowded room where you might want to listen to different conversations, and thus need no single conversation to be too loud compared to the others.

Over the air broadcasting uses both the VHF and UHF frequency bands. This is true for both analog and digital broadcasting. As a result, there are three broad classes of antenna used to television: those that are optimized for UHF reception (and terrible for VHF), those that are optimized for VHF, and hybrid antennas that attempt to be good for both. Why isn't every antenna a hybrid? Because if you optimize an antenna for only for one set of frequencies (for example VHF, but not UHF), you are often able to produce one that is either smaller, lighter, more selective, or cheaper than an equivalent hybrid antenna. As with people, being extremely specialized has both pros and cons.

The general rules of thumb for antennas are that they should have as clear a line of sight as possible to the transmitter sending the signals they are supposed to receive. This leads to some very simple guidelines for the placement of outdoor antennas. The rules of thumb can be expressed with the acronym **CHOB**: closer, higher, opener, bigger. This means signal reception is better for transmitters that are *closer* the source of the signals (all other factors being the same), signal reception is better when the antenna is placed *higher* up, signal reception is better then the are is more *open space* around the antenna (for example outdoors as opposed to indoors), and signal reception is better when the antenna is a *big* as possible. The same guidelines need to be taken with a grain of salt in the special cases of either indoor antennas, or situations where there may be major obstructions between yourself and the broadcast station of interest. In the special cases of indoor or obstructed environments, these rules may not always apply since while the same physics is at work, there may be other hard-too-see factors involved since the signal is probably bouncing around and, for example, getting closer to the source might cause the antenna to move off a useful bounced signal path.

Directional selectivity is an important parameter and refers to the extent to which an antenna selects signals in a preferred direction, and rejects other signals. It is sometimes referred to as **directivity**. A special case of directional selectivity is the ability to accept signals in the direction where the antenna is pointing, but reject signals in the opposite direction. This is known as the **front-to-back ratio**, and usually depends on a design involving a shield-like construction (see Figure 37).

Most television signals, especially in North America, have horizontal polarization. As a result, the elements of television antennas are always arranged to lie horizontally.

Figure 35: Compound antenna composed of many horizontal dipoles all with the same frequency.

Antenna hardware components

Antennas are constructed using a fairly small set of basic components and principles. It is useful to know the names of these since they both identify the antenna and determine its main properties.

Many antennas are composed of smaller antenna elements, arranged along a boom that supports them, but the boom does not perform any electrical function.

The dipole

A dipole is a basic concept from physics, and in the context of antennas it refers to a short stick. Many antennas use short stick-like or loop-like elements that lie across the signal path. The simplest antenna is composed of just one dipole (and is called a dipole antenna). A folded dipole is a variation where the stick it made out of a flattened loop of wire, and a bowtie dipole is a version where the stick is composed of two triangular loops (see Figure 37).

Compound antennas

By combining multiple antennas, the electrical signal from each antenna can be combined. This could be achieved by connecting completely separate antennas, but it is more reasonable to construct a single compound antenna from multiple "sub-antenna" elements. A common way to do this is the stacked dipole antenna, in which a succession of dipoles is mounted along a single boom. If all the dipoles have roughly the same properties, they can be used together to obtain better reception at a specific frequency. On the other hand, by using dipoles that have different frequency preferences, an antenna with a broader tuning can be achieved.

Log periodic dipole arrays

Log Periodic Dipole Array antennas (LPDA) are a particularly common type of compound antenna (see Figure 36). They are composed of a series of successively longer dipole "sub-antennas" mounted and equally spaced along a long boom. The advantage of the different dipole length is to allow the antenna to pull in a broad range of different frequencies. Each successive dipole is connected to the next one across the boom, but on the opposite side of the boom (i.e. the first dipole on the left is tied to the second on the right, which is connected the third on the left, etc.).

LDPA antennas tend to be fairly compact, yet they work across a broad range of frequencies. The dipoles can be simple parallel antennas, or they can be angled forward slightly (in a vee-type antenna), leading to slightly higher performance for selected channels.

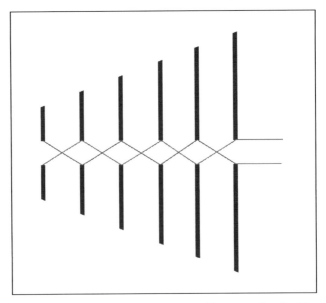

Figure 36: LPDA antenna sketch. Black vertical bars are dipoles (the narrower front end is pointed towards the transmitter).

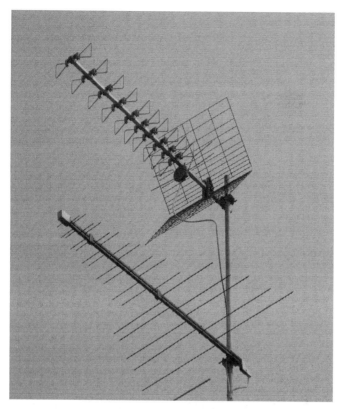

Figure 37: Compound antennas made up of horizontally aligned dipoles. The top antenna uses bowtie dipoles and provides high gain UHF reception and has a shield to improve the front-to-back ratio. The lower one is a typical LDPA antenna.

Multipath and ghosting

Multipath interference is a generic signal transmission program that occurs with both analog and digital systems. Multipath **ghosting** is a phenomenon that is generally only seen with analog video broadcasts, and thus it is included here partly as a historical note. It results from a signal bouncing off an object on the way from the transmitter to the receiver, even when there is also a clear line of sight between the two. In other words, the transmitted signal simultaneously takes two different paths to the receiver, which is why the core transmission problem is called **multipath** interference since duplicate copies of the signal interfere with each other. As a result, the receiver gets two copies of the transmitted signal, one that came directly from the transmitter and one that took a longer bounced path, and thus arrives with a slight delay due to the extra distance traveled (even at the speed of light, this matters!).

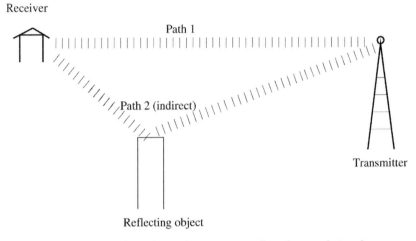

Figure 38: Multipath interference comes from bounced signals.

In the case of analog video, the receiver displays the resulting incoming signal as a combination of two time-offset versions of the same transmission. Since analog video is transmitted as a string of successive brightness values staring with the beginning of the frame, the effect is to add a second (delayed) version of the image onto the original picture, thus producing a picture with a ghostly offset duplicate added to it. In some cases, if there are several alternative bounced paths between the transmitted there can be several ghost images. Some very smart analog receivers attempt to detect the fact that the signal coming down the antenna contains a duplicated version of itself, and compensate for it.

Figure 39: Original image without ghosting (see below).

Figure 40: Example of ghosting due to multipath interference (compare with original above).

While digital signals also have the potential to bounce as they travel, and this can cause serious signal degradation, it does not look like ghosting. This is because there is enough extra information encoded in the packets to allow the ghost images to be rejected. Instead, packets may simply be rejected or else the signal is damaged so badly that nothing al all shows up. In fact, with COFDM transmission (see page 74) the existence of multiple alternative signal pathways is encouraged and exploited to assure *better* reception, on the assumption that we can pick the best signal path we see.

Indoor antennas

General guidelines for the placement and selection of indoor antennas are elusive since what works best depends on many factors specific to an individual home. For example, the thickness of the walls, the construction materials used, and the attachment to any adjacent home can all be important factors, and there are lots more to boot. If your house has metal siding, you can expect to have especially poor reception with an indoor antenna.

Even with an indoor antenna, in general, the CHOB principles should still be considered, but they may be overridden by all the other issues and devices in the home. The proximity of other electrical devices can really make a difference (usually a negative one). Unfortunately, the reception may also vary with the positions of people in the home, or other factors, so what works best at one moment for one station may not be ideal later on or for a different station. Many of us have had the experience of painstakingly adjusting an indoor antenna to get a program of interest, getting it just right, and then having the signal vanish once we sit down to watch the show. This is often because the signal is being affected by our body, either by bouncing off it, of having it act as a shield against a source of interference.

In general, an indoor antenna will provide a signal that is at least 30 per cent weaker than what would be achieved with an outdoor antenna of the same size, and indoor antennas are always much smaller than outdoor ones, so don't expect to pull in really weak stations with just an indoor antenna. As a rule, the reality is that one must tinker experimentally with the indoor antenna placement until it works as well as possible.

It may help to put a moderately long cable on the antenna so you have the freedom to hunt for the best locations. A position near a window is often good, although even that simple strategy will depend on where the transmitter is. Having the antenna face towards the transmitter is also a good rule of thumb (but will be violated if the best signal only arrives via an unpredictable bounced pathway). If the transmitter is within 10 miles and is not obstructed by a mountain or major structure, you should be able to get a good signal with just an indoor antenna. For transmitters that are ten to 20 miles away, the situation is unpredictable and the performance of an indoor antenna may be intermittent. For transmitters more than 20 miles away, an indoor antenna is not likely to be suitable.

An intermediate step between and indoor and an outdoor antenna is the use of a small outdoor antenna, but placed inside the building, either inside a room or up in an attic. Such an arrangement may result in performance that us much worse than placing the same antenna outdoors, but can be useful when an outdoor fixture is impossible. A two-bay UHF antenna for outdoor use can be as little as 30 to 60 cm (foot or two) long/ This may make it small enough to tolerate inside a room, even if used in conjunction with a VHF antenna.

Indoor antennas can be approved for use in the USA by the Consumer Electronics Association. This approval is not required for good performance, but it does assure a minimum standard of fabrication.

Outdoor antennas

While the same antennas can be used for analog and digital television, terrestrial reception of digital TV does not work at all with borderline signals. Furthermore, there the signal strength on the antenna needs to be not-too-strong, and also not-too-weak. As a result, outdoor antennas for digital television are often highly directional: they need to be aimed at the transmitter of interest (especially in North American where VSM instead of

COFDM transmitting is used). This, in turn, implies that it is often desirable to have either multiple antennas, or to have a **rotor** attached to the antenna so that it can be mechanically turned when the station is changed.

Using a preamplifier

If the signal coming in is weak, an antenna **preamplifier** between the antenna itself and the receiver can be used. It is used to boost the signal strength from an antenna before it reaches a receiver. Do not use a preamplifier if the signal is already strong since then it can overload the receiver and make reception worse. In general, a preamplifier will boost both the signal and any undesirable electrical noise as well. Thus, a preamplifier will make the antenna signal stronger, but separate the broadcast signal of interest from other electrical interference that is mixed in (which is often the case with weak signal in urban environments). As a result, a preamplifier cannot compensate fully for a weak signal, and thus if the reception of a channel reception is bad, a larger (or better placed) antenna is much more likely to lead to a good result.

In general, when a preamplifier is used one would like to amplify the weak antenna signal without allowing any additional electrical noise to creep in. Thus, the preamplifier should be physically placed as close as possible to the antenna to avoid the need for a long cable run (since a weak signal will degrade even on a good quality cable). In many cases, including my own, people buy preamplifiers and discover they don't help much, so try and borrow one first and evaluate it if you can.

A **distribution amplifier** serves a different purpose: it boosts the signal strength so a signal so that it can travel longer distances. A distribution amplifier is used when the length of the cable run, typically coming out of the receiver, is very long, or when it splits (for example to drive two different receivers).

Balanced and unbalanced cables

Sometimes one needs to connect a two-wire antenna, such as from a dipole, to a coaxial cable or coaxial jack. A coaxial cable is an unbalanced cable, since only the center wire is powered. A ribbon cable or dipole antenna is essentially a single loop, and is balanced, since the two wires play equivalent and complementary roles. A **balun** is a small electrical transformer that allows a balanced and an unbalanced cable (or pair of devices) to be connected together. The named comes from the first letters of the words "balanced" (bal) and "unbalanced" (un). A balun can also, but not necessarily, accomplish impedance matching: allowing cables or devices with different electrical impedances to be connected together. A balun is needed, for example, to connect an unbalanced coaxial cable with 75-ohm impedance to a device that expects a balanced 300-ohm ribbon cable.

Like almost all transformer, the basis of a balun is the use of a magnetic "connection" between the input and the output (known as the primary and secondary circuits, in transformer lingo). A typical balun is made up of two wires loops, one from each of the balanced and unbalanced sides, wrapped around a toroidal or cylindrical core made of ferrite.

II. Conditional Access Control & DRM

When is comes to data security for video content, there are two rather different kinds of data protection that are relevant. The first kind is access control: only allowing certain viewers to view certain channels. The second kind of data protection relates restricting what people can do with a stream of data even if they can view it, and in particular restricting their ability to copy it. Allowing viewers to watch a channel only on the condition that they have paid for it, or if they are in a specific geographic region, is called **Conditional Access**, and it is used for cable TV channels, satellite channels, and pay-TV in particular.   Restricting the ability to record a program, or make a copy of a recording, is called **Copy Protection**, or sometimes as **Digital Rights Management** (DRM).

The all-encompassing approach to the administration of restrictions on video content has been developed in the form of the Content Protection System Architecture (CPSA), developed by a consortium of large consumer-electronics companies. This architecture seeks to protect and control data use and transmission along every step in the chain from production and broadcasting to display in the home. The basic concept is that every linkage needs to be secured and restricted; typically is accomplished using encryption technologies and standardized data handling procedures encoded in the content itself, or something equivalent. The management information can be embedded in the audio or video stream, but it is most commonly embedded in the video and referred to as Content management Information

(CMI) and can be encrypted or unencrypted, and can be either digital or analog. The CPSA also seeks to control what the broadcasting industry refers to as the **analog hole** which refers to "loophole" in the encryption barrier arising from the fact that analog video is hard to control and encrypt, partly since so many well-established technologies are already in place, and compatibility with them must be maintained. In other words, the use of analog video signals provides a means for video duplications and transmission that is like a hole in a security barrier. If you can watch it, you can usually record it without too much cost or effort.

Many copy protection systems depend on embedded messages that are transmitted along with the programming. Such metadata is used to indicate how the programs can be manipulated, and what can be done with them. A simple example of such a message is the Broadcast Flag. This is a token that can be set when programs are transmitted, which specifies that they cannot be recorded (even on a PVR). A related flag is used for Selective Output Control (SOC) when, if present, tells video equipment that certain types of output connector (typically analog outputs) should be disabled. These particular flags are not in standard use, but they supported by many video devices and serve illustrate the general concept.

Alternative copy-protection technologies			
Name	*Context*	*Data type or target medium*	*Comments*
CPRM		system-wide	
APS	Transmission	Analog	Very wide use
CGMS-A	Transmission	Analog	Moderate use
HDCP	Bi-directional Transmission	Digital	Used on most recent HDTV displays
DTCP	Bi-directional Transmission	Digital	Common on recording devices
CGMS-D	Transmission	Digital	Subsystem for HDCP/DTCP
CPCM	DVB streams	Digital	
Broadcast Flag	Transmission	Digital over-the-air broadcasting	Status in flux
CSS	Media (DVD)	Digital	Used on all commercial DVDs
AACS	Media (DVD)	Digital	
VCPS	Media (DVD)	Digital	
TivoGuard	Tivo PVR devices	Digital data on Tivo Devices	
Microsoft Windows Media DRM		Computer media	
Verance		Digital audio	
Open Mobile Alliance (OMA)	Mobile device media	Digital	
Veil	Media		Watermark

Conditional access

Conditional access (CA) refers to limiting what can be watched, and how the data can be manipulated or copied. In general, it boils down to two fundamental kinds of operation: handling the encryption of the signal, and handling permissions and messages and interact with the encryption system to allow messages to be decrypted under certain circumstances.

Traditional analog television was broadcast through the air and could be received by any suitable TV set. The protocol and electronics were simple enough to make it essentially impossible to encrypt the signal, and thus there was no way to force potential recipients to pay for a signal they might wish to receive. While there are many ways to encrypt an analog signal, most of them are awkward and could never be retrofitted into the existing technology. The only form of access restriction that existed was based on geographic location: some programs might not be broadcast in certain areas.

With the advent of digital encoding, encryption becomes easy to achieve. Since a receiver needs to include a computer to decode the signal, adding an extra layer to the software to implement encryption was fairly easy. Access restriction, known as conditional access, is used to implement various kinds of fee-based service, such as pay-per-view movies, requirements for a subscription to specific channels, and control over what geographic locations can view certain channels. All digital television standards, and DVB and ATSC in particular describe methods for encrypted data streams and "conditional access" which depends on the receiver being authorized somehow. Within the high-level protocol description, there is room for individual programming providers to implement different variations of how the low level details work.

The DVB protocol includes specifications for conditional access protocols (DVB-CA), data encryption systems (DVB-CSA), and the connection of hardware smart cards (DVB-CI). These protocols allow a provider into "insert" proprietary sub-systems within a general abstract framework, for example, by calling a secret proprietary encryption using a standard function-calling mechanism. One standard high-level architecture for Conditional Access Modules (CAMs) is used for all encrypted DVB programming. For stand-alone receivers, these generally take the form of a slot into which a smart card is inserted. For computer-based systems, the tuner card usually includes a conditional access module (CAM), which is a card reading that the smart card can be inserted into.

Since many users would like to obtain programming that they may not be authorized for, there is a continual battle between the service providers (and the contractors that provide CA systems to them), and the user community (especially those known as "pirates"). In general, almost all CA systems that are in widespread use are eventually "cracked" and methods to circumvent them become available to the public.

For many reasons, the rise of access controls on programming has met with some resistance. That is, users have developed technologies and techniques for circumvent the access controls put in place by broadcasters. Some of this resistance is based on idealistic principles regarding the need for free information or the concentration of power in a few companies, some is based on a desire to escape geographic restrictions on available content, much is based on curiosity-driven hobbyist motives, and a substantial amount seems to be driven for a simple desire to obtain programming for lower cost. All these motivational forces have acted in concert to lead to a small, but rather vigorous, ongoing background of activity aimed at circumventing conditional access methods.

There are some who believe that a limited degree of CA circumvention (i.e. media "piracy") is actually beneficial to media producers and distributors. To oversimplify the rather complex arguments, the result of media piracy is often that the content gets more widely distributed and exposed, but in an inconvenient form. As a result, the content being delivered gets positive exposure and when the pirates become sufficiently fond of it, or become sufficiently wealthy, they are converted into legitimate customers to avoid the inconveniences. The validity of this argument depends on a range on unverifiable factors, such as the rate that media pirates are converted to genuine customers, and how many of them might have purchased the content in the first place.

Smart cards for access management

A key link in the conditional access chain is the smart card. This is a small processing device that physically resembles either a credit card or a telephone SIM card, and which usually typically resides inside a side door of a device. A smart card processes messages from the broadcaster and provides encryption keys or other control information to the set top box that control its behavior.

All smart cards for television access control are based on the ISO-7810 and ISO-7816 international standards that define their physical characteristics like size and shape, as well as their electrical needs and behavior. The same standards are used for telephone SIM cards and other kinds of security device. The cards use a serial interface for communications and typically include a microprocessor, various types of volatile and non-volatile (i.e. permanent) memory, and get power from the host they are connected to.

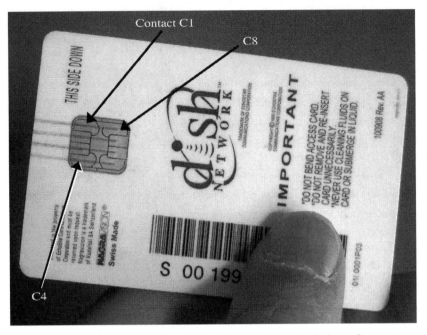

Figure 41: Example of a smart card. The contact pad divided into 8 separate contacts can be seen on the center left.

A small contact area near one end of an ISO-7816 card is divided into several connectors for data, clock, reset pin, and power. The pins are numbers C1 through C8 moving counter-clockwise from the top-right, as listed below. When the reset pin is grounded briefly, this resets the card. On being reset, every card responds with a fixed series of byte values called an ATR (Answer to Reset) and this allows any card model to be identified. The first character of the ATR (the TS byte) indicates what kinds of communications are used, for example whether high voltages represent Boolean true bits or false bits. The second byte provides information regarding how much additional information should be expected in the rest of the ATR message. Additional bytes specify the manufacturer and other information. For example, an ATR response code of

3F 67 25 00 26 14 00 20 68 90 00

is indicative of the Pay-TV card from Casema Cable Television of the Netherlands and specifies a 5 volt programming voltage.

Smart card contacts		
Position	**Technical**	**Abbreviation**
C1	Vcc	Supply voltage
C2	RST	Reset pin
C3	CLK	Clock frequency
C4	RFU	Undefined/Reserved
C5	GND	Ground
C6	Vpp	External programming voltage
C7	I/O	Serial input/output communications
C8	RFU	Undefined/Reserved

Commonplace smart cards used for cable or satellite television authentication often use specialized microprocessors. A recent model of smart card for Dish Network, for example (the "ROM102 card"), apparently uses the special-purpose ST19XL microcontroller (made by the Dutch company STMicroelectronics), a variant of the more generic ST7, which in turn is built around a standard ARM microprocessor core. Although STMicroelectronics is not a company familiar to much of the lay public, they are one of the largest semiconductor companies in the world. STMicroelectronics microcontrollers are, in fact, a leading family used in embedded applications including not only smart cards, but also telephone SIM cards, pay-tv cards, digital receivers, and diverse embedded systems and related security applications. Since smart cards are regularly used for enforcing security policies, their construction and internal design is often not released to be public. The ST19 processor family, for example, is not accompanied by public documentation, unlike the more commonplace ST7 that it is based upon. In fact, there is an ongoing cat-and-mouse game between manufacturers who include hidden features in these systems, and a wider community that tries to reverse engineers them for various reasons.

As with any RAM-based computer, part of the memory is reserved for the execution-time stack, and part for global variables. A standard feature of many smart cards is the use of programmable ROM memory (PROM) which can be modified repeatedly, or in some cases only once. One-Time Programmable ROM (OTPROM, or just OTP) can sometimes be changed by program execution, but once changed it cannot be restored to its original value.

Electrically Erasable Programmable Read-only memory (EEPROM) functions somewhat like a hard disk. It can be used to store program code, authorization codes, or other data, and this data is not erased when power is removed. EEPROM memory functionality can be implemented with flash memory (technically, EEPROM and flash memory are based on different technologies, but they both provide non-volatile storage). Typically, some portions of the executable code for a smart card is located in EEPROM memory. This allows broadcasters to change the function of the card, or the encryption protocols, by transmitting new program code that over-writes some parts of the EEPROM thus changing the operation of the card (and hence the device that contains it). This can extend all the way to completely disabling a card if the broadcaster can determine that the code inside it is somehow incorrect (using checksums or other methods).

A well-known example of an esoteric use of one-time programmable (OTP) memory occurred in 2001 one week before the very popular "Superbowl XXXV" football game. The US satellite broadcaster DirecTV used smart cards that run programs which included a self-update capability, partly as a countermeasure against users who had been successfully deactivating the existing access control code. The broadcaster had been sending small program fragments to the receivers for their satellite broadcasting service, building up to a series of about 63 successive and mysterious updates to the cards in these devices over a substantial period of time. Eventually on the night of Sunday January 21st a final update was transmitted that activated the individual fragments leading them to verify the content of the card and take decisive action. The cards whose internal serial numbers had been modified by signal pirates permanently damaged themselves. The damage was inflicted by writing anomalous bytes to the OTP memory of the cards. This particular modification included the string "GAME OVER" written over the first portion of the OTP memory block that included data needed for bootstrapping the card. The disabled "H"-series smart cards that were damaged in this way were referred to as "ice scrapers" since they were useful for little else. They also became known in the data piracy community as Black Sunday cards, although the pirates eventually found ways to reprogram the cards to work around the damage and return the cards to partial utility. In the interim, large numbers, some say tens of thousands, of cards were rendered useless, seriously crippling the television piracy community that had exploited them.

ROM memory is generally not modifiable and is programmed only when the card is manufactured. Substantial amounts of program code typically reside in ROM memory, but smart card programs often jump back and forth between program segments in ROM and EEPROM so that modifications can be implemented. The hobbyist community sometimes refers to the ROM memory in a smart card that holds executable program code as UROM, although this term is more properly used to refer to specialized ROM memory that stores microcode that allows certain types of microprocessors to change their actual instruction set.

Modification of the programming on smart card itself seems to be a CA circumvention method that involves the least ongoing hardware commitment by a pirate (once the modification has been made, often by a specialist). The piracy community refers this to as "using plastic", since the smart card itself appears to be a simple plastic card like a credit card. Subverting the operation of a card is based on changing the stored data or instructions, so that the card no longer places restrictions on available programming. Card manufacturers resist this by making it difficult to read or modify certain sections of memory. Using program code that tests the validity of the stored program is also a natural defense against tampering. In response, pirates respond with measures to make their modifications hard to detect, or by including code to ignore commands that would let the card verify itself (such anti-verification code is sometimes known as a **blocker**). Of course, the most important protection used by broadcasters is for the stored program not to accept unauthorized modifications requests, and to make the modification process cryptographically secure. Perhaps surprisingly, it has been very difficult to achieve this security in practice.

Unauthorized modification of the card has sometimes depended on persuading the embedded microprocessor to enter an abnormal state, often by "**glitching**" the voltages being used (setting them to incorrect values, sometimes for very brief intervals). From such an abnormal state, various changes can potentially be made to the card to override its protections. As a result, the embedded electronics in some modern smart cards include specific circuitry to detect abnormal voltages and shut down operation appropriately.

The abnormal alternations of the card's behavior that are exploited by pirates can readily lead to either incorrect operation or defective program code being executed; in fact this is the whole point. Sometimes, this may make the card totally unresponsive, and if it also results in modifications to EPROM, these changes can be semi-permanent. This phenomenon is known informally as **looping** the card (since an unresponsive card can be the result of the internal program entering an infinite loop). Depending on the circumstances, such damage may be irreversible.

Since STMicroelectronics is the leading manufacturer of smartcard processors and set-top box, its devices serve as a good example of smart card processor architecture. Many of these share a common microprocessor core and it is illustrative to consider the ST19 as an example. The ST19 is a large series of CPUs that are based around an 8-bit microprocessor core with a half-dozen registers. The microprocessor is comparatively slow compared to a modern desktop computer. Since they are heavily used in security applications involving encryption, however, they include a specialized hardware accelerator specifically for the Data Encryption Standard (DES) algorithm commonly used in security applications, as well as a 1088-bit modular arithmetic processor (MAP) and one or more random number generators. The MAP allows various arithmetic computations to be computed on long operands, up to up to 2176 bits in length, much more quickly than would be possible on the native 8-bit processor. The MAP and Enhanced DES (EDES) accelerator in combination can be used to rapidly perform various cryptographic algorithms.

Since smart card applications need to react robustly to power loss, both accidental and intentional, this processor family includes voltage sensors and clock frequency sensors. When the voltage gets weird, the card can shut down gracefully. Processors in the family include some with as much as 8K of RAM memory, over 60K of EEPROM (which can include a hundred of so bytes of one-time-programmable memory), and several hundred kilobytes of ROM. As a security feature, the processor can be flexibly configured with a set of rules that conditionally restrict what regions of memory can be accessed and copied to other parts of the architecture. In addition, there is even a commonplace allowance for custom circuits known as application specific integrated circuits (ASICs).

Several different access management systems based on smart cards are in widespread uses for television conditional access control. Such an approach is particularly prevalent for the reception of direct to home satellite signals. Since the number of satellite-based broadcasters is fairly small, the number of access management systems is correspondingly limited.

In addition to tampering with the cards directly, pirates may attack intercepting and tampering with traffic between the card and the receiver, by modification of the receiver, or by mimicking the card's behavior using an external computer. One exotic approach is based emulation, in which a regular computer is used to replace the smart card. Emulation attacks are based on decoding or otherwise obtaining some of all of the program stored on the card (or at least its functional description) and duplicating most of its behavior, less the security restrictions. This can be accomplished using a traditional computer connected to the receiver via the card slot, or via an embedded microprocessor that essentially constitutes a replacement custom smart card. These custom embedded emulators sometime use microprocessors based on the same ARM architecture family that is used as the core of genuine smart cards. The Atmel AVR microprocessor family is also a common choice for these devices due to its availability and good power/performance tradeoff. To defend against such attacks, one can try to make the communications between the smart card and the receiver as secure and impenetrable as possible.

Conditional access management systems

Some of the major conditional access systems in use for DVB and ATSC broadcasting include: Nagravision, Conax, Irdeto, Betacrypt, Cryptoworks, and DirecTV. All of these are built around the DVB protocols and thus have similar system architectures, and all but DirecTV are known to employ

DVB/ATSC packets in roughly the same way. Generic approaches to system security combine efforts to make the data secure, as well as procedures to make unauthorized decryption annoying and inconvenient if complete security is uneconomical or unachievable. The secret to secure access control is (a) to send instructions that restrict how video can be used, and (b) to scramble the video so that only sophisticated intelligent devices (that presumably obey the access control policies) can descramble it. In order to make the scrambling resistant to unlicensed decoding, important aspects of the decoding process are varied on a constantly changing basis, and an elaborate chain of encryption schemes is used to protect this information.

These systems are based on sending video data that has been scrambled, and which the receiver must descramble. The algorithm for descrambling is fixed, but it depends on a **control word,** essentially a password, to make it operate. This control word is a part of the decoding process that changes often (typically every few seconds). The actual decoding key is generated from data that is sent to the receiver, and which must be combined with stored cryptographic keys to produce the control word. Thus, decoding the signal depends on cryptographically combining an internally stored key and the regularly transmitted additional transmission key that is sent in a data package called an **Entitlement Control Message** (ECM). A further set of messages that specify which specific programs or channels may be decoded (and thus when the descrambling should take place) may also be sent, in the form of **Entitlement Management Messages** (EMMs). As a program stream arrives, if the EMMs indicate that the receiver is authorized to display it, only then is the key from the ECM extracted and used in combination with the private key already present. The EMM's can contain id numbers of blocks of receivers, and sometimes even identify an individual receiver, and they are generated from the subscription database of the provider. The precise way in which these various messages are formatted and used depends on the specific conditional access systems being used (e.g. CableCARD, Idreto, Nagravision, Conax, or VideoGuard).

Since no system is impregnable, the maintenance of security is an ongoing challenge for most broadcasters. As methods for circumventing their access controls become widely known, they are modified and strengthened and in ongoing cycle. These anti-piracy efforts can occasionally cause inconvenience for regular viewers ranging from software problems to the need to upgrade hardware components such as the smart card. In some regions, it may be possible to subscribe to programming from multiple providers if they use mutually compatible transmission and CA systems. Some of the major developers of conditional access and encryption systems are indicated below.

Nagravision

Nagravision is a company what produces conditional access systems of the same name, used for video programming. Nagravision is itself a subsidiary of the Swiss Kudelski Group. The access and encryption systems are widely used by many broadcasters including Dish Network and the Canadian company Bell ExpressVu, as well as Virgin Media UK and Digital+ in Spain. Due to its widespread use, it has been a very attractive target to those attempting to circumvent conditional access restrictions and thus it has been compromised many times, leading to a constant cat and mouse game between Nagravision and those who seek to circumvent its protections. At one time such circumvention was largely the province of home hobbyists, but a large commercial market for retrofitting legally-sold FTA receivers to receive and decode encrypted signals without authorization has developed, leading what appears to be commercial CA circumvention businesses.

Nagravision access control systems include both an older little-used system called **Nagravision 1**, and a more modern system called **Nagravision 2** based on more secure cryptography.

Due to the intense interest in smart card modification from such a large user base, successively more secure access card families have been released and then named informally by the user or signal piracy communities. These names are typically based on serial numbers on the cards. Those used by Dish Network and Bell ExpressVu have been known successively as ROM2, ROM3, ROM10, ROM11, used with Nagravision 1 encryption, and ROM101, ROM102, ROM103 used with Nagravision 2.

Conax

Conax refers to a range of security products including a conditional access system developed by the Norwegian company Telenor. Their security systems, based on smart card technologies, are used in a range of counties in both Europe and Asia for all types of broadcast television. This includes a large presence in China, an unusual market for a foreign data security company.

Conax solutions are generally based on open standards, such as DVB and OpenCable, and they are among the first to have produced CA systems based on standard smart cards.

Viaccess

This encryption system was developed by France Telecom and has gone through several increasingly secure revisions. These are broadly known as **Viaccess 1** and **Viaccess 2**, although the latter includes several variants (version PC2.3 through PC2.6, and PC 3). It is used to encrypt direct-to-home satellite broadcasts, as well as transmission to cable headends, particularly in Europe.

Irdeto

The **Irdeto** encryption system, developed by the Dutch company of the same name, is widely used for broadcasts in numerous countries in Europe, Africa, China, Australia, and the Americas. Notable examples appear to include Globecast, Galaxy Satellite Broadcasting Ltd (China), China Broadcasting Film Television Satellite Co., Canal D, Showtime Arabia, Telia (Sweden) and Foxtel (Australia).

Systems based on this technique are used for cable, satellite, IPTV and other technologies. The original Irdeto system was cracked and in 2004 the CA systems was upgraded and Irdeto 2 was released. It too was compromised in 2007.

In 2006 Irdeto acquired Cryptoworks, another maker of conditional access systems, and now operates and supports its products.

Videoguard

Videoguard is a conditional access technology developed by NDS, a data security company owned by News Corp. The technology is used by News Corp's affiliated broadcasters, including the satellite broadcast companies DirecTV for their DSS protocol, and BSkyB, as well as various cable television broadcasters in the USA (see also the section on DirecTV on page 226). This technology was once the target of intense piracy. Their successive models of smart card, known informally as the "F-card", the "H-card" and the "HU-card" were repeatedly attacked and successively replaced, but are still common terms on the Internet. The eventual development of their "P4" smart card technology apparently led to many signal pirates to migrate to other more vulnerable systems.

Due to proprietary and non-standard nature of the protocols used, Videoguard access and encryption is not well supported by 3rd party manufacturers and receivers, other than those directly affiliated with News Corp. companies.

Videoguard has recently been extended from a system for conditional access for broadcasting to a system that can also be used for protection of digital media stored on disk. This merges CA functionality with broader digital rights management and copy protection. This usage model is intended to allow content to be stored on a local disk but, for example, incur a fee when it is eventually watched.

Mediaguard

Mediaguard was developed by members of the French Canal+ Group, who are its primary users. The company was originally known as SECA (Société Européenne de Contrôle d'Accès). Like many such systems, it has been cracked in the past (and subsequently re-secured).

There is a persistent rumor that a major security breach of this encryption system in the late 1990s was developed and released by the makers of a competing encryption technology. Breaking a competitor's encryption system would be a credible, albeit possibly unethical, business practice for a security company, and it would also be illegal in many jurisdictions today. The basis of this rumor may be the fact that CanalPlus launched a one billion dollar lawsuit against NDS, who provided encryption for News Corp., alleging that they had spent millions of dollars subverting the Mediaguard encryption and then releasing the results via a file called secarom.zip on a Canadian web site (DR7.com). That lawsuit was eventually dropped.

DigiCipher 2

Digicipher 2 is the encryption technique used by the broadcast system of the same name and employed on many cable TV networks as well as a range of satellite systems. It replaces the older **VideoCipher** and **Digicipher 1** technologies which were less flexible, and whose security appears to have been compromised. The **VideoCipher II+ RS** from the same organization remains in use for analog content using C-band satellite transmission. This encryption system is used, in particular, for broadcasts by cable companies to their own headend facilities.

PowerKEY

PowerKey is licensed by Scientific Atlanta for use in a variety of networking applications including cable television. It is used on the

PowerVu line of receivers, and thus the names of the encryption system and receiver are often interchanged. Depending on the application, it can be used with either a smart card or a CableCARD. It is based on a combination of private and public key cryptography and uses the RSA encryption algorithm at its core. This encryption system appears to be used primarily for commercial applications and, along with Digicipher 2 in particular, for broadcasts by cable companies to their headend facilities.

A conditional access module (software emulator) was reverse-engineered for this system and is reportedly available with some third party receivers. This is reputed to allow legitimate subscribed cards and subscriptions to be used with these 3rd party devices, but this is reportedly imperfect and is not supported or endorsed by the manufacturers. One example of such a set top box or IRD is the Linux-based DreamBox from Dream Multimedia.

Basic interoperable scrambling system

The **Basic Interoperable Scrambling System** (BISS) is an encryption system for MPEG-2 content approved for use in the European Union for various purposes including sporadic broadcasts (i.e. wild feeds) and continuous broadcasts by satellite. BISS operates in 3 modes: with unencrypted data, with encrypted data using a scrambled session word (BISS mode 1) and BISS with Encrypted session keys (BISS-E). Both mode 1 and BISS-E use the Data Encryption Standard algorithm (DES) at the core of their operation. BISS is based on a session word (i.e. a password) being entered at both the transmitter and receiver, presumably by hand. In mode 1, the 12-digit hexadecimal number must be explicitly communicated to all participants in the communication. With BISS-E, the session word that is passed around is combined with a hidden key inside the device.

BISS keys are sometimes exchanged openly on the Internet, but the source of these keys is unknown and it unclear if they are leaks, intentionally distributed keys, or if the key generation algorithm was compromised.

It is based on the DVB-CSA protocol, and thus can be used, in principle, with any DVB stream, but the standard Conditional Access table for BISS is always empty, and no EMM messages are ever used or sent.

Cable-specific conditional access

CableCARD

The **CableCARD** is a modular self-contained device for providing access control for cable television broadcasts in the United States. The idea is that this card fits into a television set and avoids the need for a separate set-top box to receive encrypted television programming. It can also be used to configure a generic standard set top box or other media device to operate with the conditional access system of a particular broadcaster. The CableCARD itself is a PC-Card (PCMCIA) device about the size of a thick credit card. Like all PC-Card devices, it has about the same electrical interface as a standard ATA hard disk; it also includes a microcomputer that governs its operation.

The CableCARD's job, to a large extent, is to convert the encryption control information of some proprietary system to a standard supported by all CableCARD devices (as defined by CableLabs Inc.). This standard is based on the Data Encryption Standard (DES) and documented in the CableCARD Copy Protection 2.0 (CCCP2.0) specification. At the time of this writing, it is supported only by cable broadcasters in the USA, but Canadian broadcasters are also discussing its adoption.

Figure 42: A CableCARD.

The CableCARD standard, although fairly new, has already gone through some revisions. The original CableCARD specification (version 1.0) only allows downstream messages to be decoded into the TV, but does not allow any messages to be sent back. This means it does not work with video on demand services where programming must be specifically requested. It also does not support the receipt of more than one program at a time (which is needed to watch one show and record another, or to allow picture-in-picture functionality). Newer **M-Card** CableCARDs support multiple simultaneous data streams, whereas the single stream versions are referred to as **S-Cards**. More recent CableCARD (CableCARD 2.0) receiver specifications allow for bi-directional data transfer while using the same kind of PC-Card-formatted devices. This permits pay-per-view requests, as well as the use of multistream interfaces that simultaneously receive multiple programs at the same time.

Transmission of information back from the receiver to the broadcaster is a function of the overall device (i.e. the television) and not the CableCARD itself. There are three different protocols that can be used for this return data stream: Aloha, DAVIC, and DSG. Each of these protocols is governed by a different standard, and they are used respectively by receivers from Motorola, Scientific Atlanta, or a mixed group of other manufacturers. In general, for a device to officially support two-way services it is expected to work with all three of these protocols. Devices that are certified as merely "Digital Cable Ready (DCR)" (an FCC certification), but which are not also

OpenCable Host 2.0 compliant, probably do not support these bi-directional protocols (note that OpenCable was previously known as OCAP).

The CableCARD operates by using a cryptographic combination of the key in the CableCARD and the key in the receiver (television or set top box) to decode messages from the cable broadcaster. Messages from the broadcasters are specific to the combination of card and receiver, which is registered during the card activation process. Conditional Access packets containing Entitlement Management Messages (as described above) instruct the receiver regarding which programming is authorized. This requires only one-way communication from the broadcaster to the receiver. For authorized programming, the ECM messages are used to accomplish decryption.

Devices that seek to support and interact with a CableCARD must be certified as being secure by CableLabs. This means that they must provide cryptographic security and digital rights management rules, and cannot be tampered with.

Most cable piracy activities seem to be aimed allowing one subscription to be shared by multiple receivers. This seems to be done by making copies of the EEPROM that stored the digital ID number of the receiver, thus "cloning" its identity, so that the duplicate receiver gains the access rights of the clone source.

Two different systems are used for conditional access and encryption in CableCARDs: PowerKey and DigiCypher 2.

Downloadable Conditional Access System

The downloadable conditional access system (DCAS) is a scheme being promoted by the cable television industry in the United States. It is premised on the use of a downloadable software system that would replace the use of a hardware CableCARD or set top box, and which would function in concert with an embedded hardware subsystem build into a receiver.

Analog video

While DRM systems to restrict the duplication of digital content are quite effective, almost all digital receivers also provide analog output. This means the analog signal (the **analog hole**) can be used to record programming. Two main techniques have been used to plug this hole: limitations on the quality of analog output, discussed earlier, and analog copy protection mechanisms.

Macrovision analog protection system (APS)

Digital television makes copy protection feasible, but as we observed earlier even with analog video there has always been a natural desire to prohibit recording. In particular, there is a strong commercial incentive to prevent either consumers or commercial ventures for obtaining commercial pre-recorded videotapes and then duplicating them (either for home use of mass resale). Several schemes for the protection of analog NTSC and PAL video have been devised, but most are not used heavily and only Macrovision technology (which is common) and Dwight Cavendish Systems (DCS) Rightsmaster technologies (which seems to be used infrequently at present) are currently prevalent. The problem is that as video recorders got better and better, they became able to record almost anything that a TV could display (which was the whole point). Anything that interfered with the video recording process would probably also interfere with the display on television. The challenge for those devising copy protection schemes was to come up with something that interfered with recorders, but not with display devices. The company called Macrovision provides two solutions referred to as Analog Copy Protection Systems (referred to by Macrovision as ACP, but more commonly known as APS, and sometimes called Copyguard) that manage to achieve this objective, although they also depend on the cooperation of the television manufacturers and especially the recorder companies to keep on building products that don't defeat this simple copy protection. These two methods can be referred to as **AGC-based**, and **Colorstripe-based**.

One of the two Macrovision approaches takes advantage of two attributes of analog television (discussed on page 15). First, as the electron gun on a cathode ray tube (old TV set) sweeps across the screen, there is an interval as it is finishing one sweep and before it starts the next where the TV set ignores the incoming signal (the vertical blanking interval). Second, video recording devices compute the average signal level to adjust the picture to give a good average brightness (this is automatic gain control, **AGC**). Video recording devices, and VHS video tape recorders in particular, use the intensity of the whole signal to compute the picture brightness adjustment, including the signal strength during vertical blanking interval. A TV set, however, does not display the content of the vertical blanking interval and essentially ignores it (with respect to brightness).

Macrovision noticed how televisions and recorders used the signal differently and drove a copy-protection wedge into this crack. Macrovision AGC-based copy protection sticks a whopping big signal (in the form of pseudo-sync pulses) into the vertical blanking interval. This causes a recorder to make in incorrect adjustment to the signal level. The effect of this is that the actual recorded on-screen picture becomes invisible, thus making the recorded copy useless. Some devices like DVD recorders, 8mm decks and the core chipsets used to manufacture them do not compute AGC in this way and, in principle, should be immune to Macrovision signals. In most cases, however, they have been equipped with extra circuits designed specifically to detect the copy protection signal used for Macrovision, and thus will explicitly refuse to record Macrovision-encoded signals even though their signal processing circuits would otherwise make it possible.

The other commonplace Macrovision copy protection technique used for analog video is **Colorstripe**. It is similar to the AGC technique and is often used in conjunction with it. It adds a corrupting signal to the color-coding ("color burst") signal that accompanies analog video (recall from earlier in the book that analog video is usually based on a separate black-and-white picture and a supplementary color signal). There are two slightly different variations of the Colorstripe system based on where exactly the corruption is inserted. Both are based on altering the phase of the color burst signal to confuse recorders and produce recordings that are marred by unappealing color fluctuations and stripes. Unfortunately, many sources claim that

Macrovision protection also has deleterious effects on video displays even when no copying is attempted.

The problems with this form of copy protection is that some televisions work a bit too much like VCRs, and end up having trouble with the display of these Macrovision encoded signals. A problem can also occur since some users route the signal from one device, such as a DVD player, through another device (such as a VCR) on the way to the television. If the VCR has trouble with the signal, the TV may never get a clean image, even though no copy was being made. DVD players often insert Macrovision protection into the signal artificially when they generate an output signal. Doing this properly for 480p (progressive scan) data has proven to be technically challenging.

Some VCRs, including very old ones, don't have AGC and so are immune to this kind of copy protection. It's also possible for a home user to make a small modification to a VCR, typically changing one capacitor, to circumvent some types of Macrovision protection. Locating the capacitor is quite difficult even for a suitably skilled home hobbyist. In addition, there are various specialized devices that can be used to remove the AGC part of the Macrovision protection, and occasionally the Colorstripe (for fancier devices). In addition, time-base correctors that are used to clean up professional broadcast video also remove Macrovision (and CGMS) as a side effect. Professional time base correctors are costly and exotic, but special anti-Macrovision devices ("stabilizers") have traditionally been quite affordable, costing around $100. On the other hand, since the introduction of the Digital Millennium Copyright Act (DMCA) in the USA, the sale of devices made specifically to circumvent copy protection is now illegal (in the USA, under DMCA Section 1201). This is also true for an increasing number of other countries that are adopting similar legislation. As a result, these copy protection methods are quite effective despite the fact that they can be circumvented technologically.

Although Macrovision was designed to inhibit recording on VCR devices, it is also implemented in many DVD recorders. As noted above, many recording devices have specific circuits to detect Macrovision signals, and will refuse to record an encoded signal even though the device would be capable of capturing a clean signal. In terms of signal generation, Macrovision is also produced on the analog output of many digital devices. In particular, DVD recordings can have flag bits in the MPEG data stream that enable Macrovision generation circuits in the DVD output. The inclusion and support of this Macrovision feature is a requirement of the license for the DVD CSS decoding algorithm, as well as DCTP and related technologies, and thus it is present in essentially all commercial DVD players.

The Dwight Cavendish system

The Dwight Cavendish System uses a combination of a "primary watermark" (PW) inserted into the audio signal that can be detected by signal processing systems, and a "copy label" (CL) that is inserted into the video signal that indicates what activities are permitted (copy once or not at all). The primary watermark is never modified once inserted, but the CL is designed to be modifiable. The markers, PW and CL, are used to store a random 64-bit number and a digital signature based on it (which is referred to as a one-way encrypted version). Depending on the correspondence between them, a copy of the data is permitted or not. If a copy is permitted and performed, the CL identifier is modified and re-encrypted so that it no longer matches the PW in the same way, thus inhibiting further duplication.

Much like Macrovision, DCS also inhibits duplication of the video by legacy equipment by interfering with the automatic gain control circuitry. This is achieved by injecting three types of interference into the vertical blanking interval and the synchronization pulses within it. The extent to which this might remain secure in the face of circumvention attempts, such as those used against Macrovision, is unclear.

Content Generation Management System (CGMS)

The Content Generation Management System for Analog system (CGMS-A) is also an approach for protecting analog video by encoding supplementary information in the vertical blanking portion of the signal. Rather than encode out-of-specification signals that try to confuse recorders, CGMS-A depends on the recorder being intentionally engineered to comply with CMGS directives, just the way DVD recorders do with Macrovision. The purpose of CGMS-A is to transmit the permissions that accompany the content, not to enforce them automatically. CMGS -A encodes 2-bit digital messages in the otherwise-unused vertical blanking portion of an analog video signal, and these messages specify the status of a program and whether it can be recorded.

Support for CGMS-A is a requirement for the licensing of several important video technologies, such as the CSS decoding algorithm for DVD recordings. As a result, CGMS is supported and enforced by an increasing number of recent video devices. Since the device needs to specifically detect and copy content according to CGMS directives, older devices developed before CGMS was developed certainly will not observe the restrictions. Likewise, devices that re-encode the signal somehow may accidentally or intentionally eliminate the CGMS-A signal, since it is not part of the original "traditional" specification for a clean video signal. Such devices may include low-cost video filtering devices, video scalers and, almost certainly, time-base correctors (TBCs). In general, removal of CGMS-A flags is fairly simple, but alteration of the flags without removing them is much more complex. As a result, a reasonable direction for future devices would be to somehow require CGMS-A flags to be present before processing content, although doing this for all content would prevent the device from working with old material.

The CGMS-A bit flags are encoded on different lines of the image, depending on the resolution of the picture. The details and applicable video standards are as follows.

Scan lines used for CGMS data		
Scan line on which data is stored.	Video format	Standard name and standards document
41	480p	CGMS-A+RC, Analog component video, CEA-805A
24	720p	CGMS-A+RC, Analog component video, CEA-805A
19	1080i	CGMS-A+RC, Analog component video, CEA-805A
20	525i and 480p	CGMS-A, IEC 61880
41	525p 480p	CGMS-A, IEC 61880-2
21	525i 480i	CGMS-A, EIA/CEA-608-B

The protocol called **CGMS-A Redistribution Control (CGMS-A+RC)** refers to a more recent (year 2003) extension to CGMS-A that also specifies if content can be redistributed in various ways, such as over the Internet. It was accepted as a standard by the CEA as part of a data packet called Redistribution Control Information (RCI) and is used for with 480p, 720p and 1080i signals, in particular.

Digital Video Copy Protection

In general the protection of digitally encoded video is much more secure than analog video. This is both because there are fewer legacy systems to support, as well as because digital data is more easily manipulated. Part of the vision of media producers is to build and require a fully-secure set of links leading from media production to consumer display: so-called end-to-end content protection.

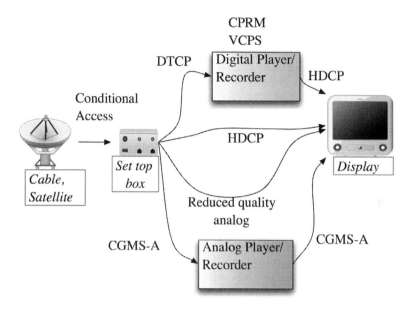

Figure 43: End-to-end content protection.

Content Protection and Copy Management (CPCM)

CPCM is an extension to the DVB protocol to allow for control of data duplication and distribution. Like almost all important protocols, this one is evolving. It allows for Usage State Information (USI) as part of the DVB signal that indicates how it can be used. It deals specifically with five kinds of usage: 1) copy and movement control; 2) consumption control; 3) propagation control; 4) output control; and 5) ancillary control. In principle, it allows for very diverse and specific types of variation in how different kinds of material can be used. Since it is retrofitted onto existing DVB protocols, it requires that receiving equipment recognize the codes and perform accordingly. The guarantee that restrictions are observed in practice is typically achieved using a combination of licensing restrictions and legislation.

Content Scramble System (CSS)

The Content Scramble System (CSS) is the encryption technology used to protect commercially copy-protected DVD recordings from being duplicated. As such, it has been is use for some time and is the basis for several more advanced systems that seek to improve upon its weaknesses. The CSS decryption algorithm is legally protected and can only be incorporated in commercial devices under license from the DVD Copy Control Association. A condition of this license is it that the device must prevent duplication of CSS-protected material.

Although the algorithm was broken and the **DeCSS** source code for decrypting CSS protected material is widely available, CSS still prevents many users from duplicating DVD material. Legal restrictions prevent commercial applications from using the widely available DeCSS code even though the code itself is available as open source. As a result, DeCSS code is generally available only as non-commercial software for home computers. Many users are unwilling or unable to seek out programs that provide the ability to duplicate DVDs, concerned over uncertain legalities, or are unmotivated to bother with the cumbersome duplication process. On the other hand, the widespread availability of the DeCSS source has permitted various non-commercial projects such as Linux to provide DVD playing software where a commercial license would have been impossible. This simple desire for a Linux-based player was, in fact, the impetus behind the initial successful effort to develop DeCSS. The dispute over the legality of distributing the DeCSS code has become something of a *cause celebré* in the open source community, leading to people translating the source code into different computer languages as well as (recoverably) into bar codes, audio files and other bizarre forms. In addition, another program with a different and non-controversial function, also called DeCSS, has been developed and it widely distributed, apparently in part just to complicate the efforts of those who try to find and shut down sites offering DVD-related DeCSS code.

A supplement to CSS that was used to prevent disk duplication is called ARccOS. This protection system, developed by the Sony Corp., is based on the introduction of bad sectors into a DVD in order to confound duplication programs. Unfortunately, it apparently also confuses some commercial DVD players, while software for DVD duplication found ways to work around it. Presumably as a result of this, the system was discontinued, although DVDs protected by it are still in circulation. Some manufacturers seem to have migrated to alternative technologies such as **RipGuard**, from Macrovision. RipGuard alters the formatting of the DVD data and is more difficult, but not impossible, for duplicating software to circumvent. It too can reportedly render DVDs unplayable on some players.

High Definition Copy Protection (HDCP)

The **High-bandwidth Digital Content Protection** (HDCP) System is an approach for preventing duplication of digitally encoded video content. It is based on the use of digital data encryption to restrict what devices can receive content. Media (such as a DVD or a specific program being received) can set a flag called the **Image Constraint Token** (ICT) which specifies that full-resolution data is only to be provided to HDCP-compliant devices. These HDCP devices will not permit the recording of a program that bears an ICT. This procedure is intended to protect each stage of the delivery system starting with a receiver (or player) and ending with a television display device, so that recordings of a program at full-quality can be prohibited. When an HDCP source is connected to an analog output (which by definition is not HDCP-complaint), the maximum resolution available is limited to at most 960x540. When connected to a non-HDCP-complaint digital device, the transmitter will supply no signal whatsoever.

Data transmission between each major component of a system requires bi-directional authentication using data encryption techniques. The protocol was developed by the Intel Corporation and cannot be used without a license from an Intel subsidiary called "Digital Content Protection LLC". Despite this restriction, it has been integrated into various international standards and is widely adopted in consumer electronics products. In particular, it is used to control transmission along **Digital Visual Interface** (**DVI**) cables such as those that go from a computer to a monitor, **High-Definition Multimedia Interface (HDMI)** connections such as those between a receiver and a television set, Giga-bit Video Interface (GVIF), and Unified Display Interface (UDI) connections. Note that devices that support HDCP can also accept input from unencrypted sources.

The HDCP protocol is defined with respect to data sources, called transmitters, and data receivers. A transmitter, in this context, could be a DVD player or a satellite receiver, and a receiver could be an LCD monitor. Before a transmitter sends an HDCP video signal to a receiver, it engages in a two-way data exchange to assure that the receiver is authorized to receive content. It also verifies that the receiver is not blacklisted on a list of devices that might once have been authorized, but for which authorization has been removed (presumably because the manufacturer of the device somehow broke the rules). If the receiver passes these tests, then encrypted content is delivered to it using shared secret keys agreed upon during the authentication process.

The authentication process is based on the receiver and transmitter each "proving" that they internally hold genuine secret information (i.e. they both know the secret HDCP handshake). This information is based on a set of 40 secret keys, each 56 bits long, that are stored inside every HDCP-complaint device. The keys themselves are never exposed, but instead the devices exchange information about which subset of a set of 20 keys should be summed together based on **Key Selection Vectors (KSV)** that the devices send to each other. The keys are constructed so that even though two devices have different internal secret keys, they are both assured of getting the same result from the calculations with keys specified by the KSV. As an over-simplified example, one device might be using two keys with values 2 and 7, and the other might be using 4 and 5, but both devices get the same sum (9) if they ask each other what they get when their keys are added together. In short, they ask each other mathematical questions that use their keys, and thus can test each other's keys for validity without actually having to know or expose precisely what they are.

In order to get the encrypted keys used to identify a device, a manufacturer needs to satisfy Digital Content Protection LLC that the device is secure and complies with their requirements. If at some later date the device is found to be unsatisfactory (for example it gets cracked and used to decode content), its key selection vector can be added to a **Key Revocation List** that is embedded in each piece of HDCP content. If the transmitter finds a device's key on the revocation list, it will refuse to transmit a signal. Thus, as time passes we can expect new content to come with larger and larger revocation lists. The legal and commercial implications of revoking the validity of a legitimate consumer's model of HDTV device after they already own it seem unclear.

Once the transmitter is satisfied with the validity of the receiver, and both have made the required calculation to get a common value from the calculation with their secret keys, the real data transfer can begin. Each frame of video is encrypted with a block cipher before being sent, and the encoding algorithm is seeded with the common shared value. Before each frame, new encryption seeds are computed and the receiver and transmitter re-synchronize checksum-like values after every 128 frames, or every 2 seconds, whichever is less. A special additional phase of the protocol is used to allow for a repeated transfer between the transmitter and receiver.

The protocol apparently has weaknesses, found by a team of researchers at Carnegie-Mellon University, Zero-Knowledge Systems and the University of California at Berkeley (Crosby, Goldberg, Johnson, Song, and Wagner). Given 40 devices with known secret keys, these devices can conspire together to ask suitable questions and determine the keys from another 41st device. Furthermore, all HDCP encryption is based on a matrix of 1600 secret numbers: the secret keys and KSV's. Since there is already an algorithm for extracting the keys from a device, it seems inevitable that the HDCP keys will eventually become public, if they aren't already by the time you read this. On the other hand, legislation in the USA and other countries prohibiting decryption technologies make it unimportant that the encryption of HDCP is insecure, so long as the manufacturer *tried* to implement secrecy. As an aside, there do exist secure algorithms for cryptographic key exchange, such as the one originally developed by Whitfield Diffie and Martin Hellman (the eponymous Diffie-Hellman algorithm), but they were not used for HDCP. The conjecture in a 2001 paper by Crosby, Goldberg, Johnson, Song, and Wagner that first exposed the flaw in HDCP encryption is that the cost in transistors for a more secure algorithm was too high. HDCP may have had to contend with a budget of some 10,000 gates, so the current algorithm based only on addition instead of more expensive mathematical operations was, like most things in life, a compromise.

Digital Transmission Content Protection (DTCP)

The **5C DTCP** copy protection protocol (sponsored by 5 companies, Hitachi, Intel, Matsushita/Panasonic, Sony and Toshiba) was also designed to restrict duplication of HDTV content. It is used primarily for DVD content, cable television and Japanese terrestrial and satellite broadcasts. The system was developed for content delivery over a serial or serial-like bus. In fact, it was developed specifically for content transmitted over the Firewire electrical protocol (also known as **IEEE 1394**) and has been subsequently extended to other transport systems including content delivered over the Internet (DTCP-IP), USB and Bluetooth. Like HDCP, it also requires bi-directional communication for authentication and key exchanges (AKE) between the device providing the content and the television that will display it. Once authentication is completed, subsequent data transfers are encrypted. It provides for three different types of content control: programs that can be copied and recorded without restriction, programs that can be copied no more than once (called **Restricted Authentication**, or **Copy-One-Generation**), and programs that can not be copied or recorded whatsoever (called **Full Authentication**, or **Copy-Never**).   In addition, content that is duplicated from what is marked **Copy-Once** is flagged with **No-More-Copies**. This Content Control Information (CCI) is embedded in the program stream, for example as MPEG-2 packets.

In general ATSC broadcasts and cable broadcasts that are sent "in the clear" are expected to have the unrestricted setting, pay-per-view content typically has the "copy never" setting, and the "copy once" setting is to be used for other types of intermediate premium content.

DTCP was designed to inter-operate with HDCP, so that a DTCP device is willing to transmit protected content on to HDCP devices, on the assumption that copy protection rules will be obeyed, and the security of the data will remain intact. The cryptographic algorithms for DTCP are based on the M6 algorithm for Firewire and related communications, and on AES-128 for IP-based data transfer. In each case, three different cryptographic keys are used. The Authentication Key, produced by the authentication process is used to protect the Exchange Key, which is used to set up the content streams to be delivered. This, in turn, enables the use of the **Content Key**, which is used to encrypt the actual AV content being delivered.

The DTCP-IP variation of the protocol is used for data delivered over the Internet, including wired links. It takes measures to preclude widespread broadcast or retransmission of the data. This is achieved by additionally restricting both the maximum hop count of the IP packets to 3 hops, and also to restricting the maximum transmission delay to seven milliseconds.

The licensing of DTCP requires that the manufacturer of a device follow rules regarding how copy protected data is stored, cached (during a pause, for example) and saved on secondary storage, as well as how the device itself is designed and manufactured. At the time of this writing, license fees for DTCP are not public, but were unofficially rumored to be on the order of 10 to 20 thousand dollars, plus 5 to 10 cents per device.

Content Protection for Recordable Media (CPRM)

Content Protection for Recordable Media (CPRM) is an encryption and access control system for recorded media, specifically DVD-R, DVD-RW, and DVD-RAM disks and is used to permit single-copy recordings. It has also been extended to apply to pre-recorded media, in which case it is sometimes referred to as **CPPM**, as well as to SD/SDHC memory cards. It is also usable with other recording media such as memory cards, assuming they have the required ID codes embedded in them. It was developed by the **4C** consortium made up of four companies: IBM, Intel, Matsushita and Toshiba.

The system is based on encryption that exploits hidden read-only content permanently embedded in the media by the manufacturer in the form of a set of **Media Keys** (making up a media key block) and **Media ID**, combined with 16 secret Device Keys stored in the hardware player. The key used to decode the media is called the **Title Key** and it, in turn, is obtained using the **Media Unique Key** which is derived from the Media Keys in the hidden block. The media key block is a table that stores the media key repeatedly, using different alternative device keys to encrypt it.

On pre-recorded media these keys reside on the **Control Data Zone** of the **Lead-in Area** of the medium. This is a region that cannot be modified by home recorders. On recordable media, the media keys need to be stored in a device-dependent manner. For example with SD memory cards, a **protected area** or **system area** is used to store the keys. These regions are not accessible to normal user software, but only via special security commands (described in section 3 of the SD specification).

The intent with both CPPM and CPRM is that once a recording is made, it can only be played on licensed devices that comply with recording restrictions and which typically will prevent subsequent copies from being made. A device with the right keys can decode the media keys, which can be up to 65,636 rows of 16 values, and use these to decode the actual data.

The system allows for alternative cryptographic methods to be used, but in practice it uses the **Cryptomeria** (or **C2**) **one-way** algorithm based on a 56-bit key and a symmetric encryption algorithm. It operates using multiple rounds of bit shuffling accomplished using shifting and addition using a **Feistel** network, which is also used by the Data Encryption Standard (DES) and other cryptographic protocols (albeit typically with other mathematical operators). In 2004 a distributed attempt to crack this cipher was launched in Japan, where CPRM use was slated to be initiated for television broadcasts. Although all possible 56-bit keys were apparently examined, the correct key was never discovered.

CPRM, like several other modern device-based protocols we have discussed, also includes a mechanism for key revocation. Some variations of the protocol also allow for keys that can only be used for a limited time, based on the date of issue. Licensing fees for CPRM memory card technologies have been, as an example, on the order of $9,000.

Verance watermarking

The **Verance Copy Management System** (VCMS)   watermarking technology inserts inaudible data in the audio stream to mark recordings as copy protected. It can be used for both audio recordings (**VCMS/A**) and video data (**VCMS/AV**), but seems to use similar methods for either type of data (i.e. putting the watermark into the audio stream). Recording devices that detect the watermark can then refuse to record the protected content. The watermark is apparently robust to re-sampling and basic manipulation of the audio signal. This watermark carries a 72-bit payload composed of 4 copy-control bits as well as 8 bits of usage information and 60 bits to identify the particular content. The **copy control bits** (CCI) can specify that either zero, one or an unlimited number of copies are permitted. Support for the Verance watermarking technology is a requirement for obtaining a license to CPRM technology.

Video Content Protection System (VCPS)

Yet another form of video content protection is **Video Content Protection System** (VCPS). This is a system used to protect content on recordable DVD media, specifically "plus" media (DVD+R and DVD+RW). It was designed to allow protected content that was delivered over the air to be recorded on a one-time basis, but to prevent subsequent copies from being made (using the US **Broadcast Flag** that marks content as not-freely-recordable). The broad outline of how VCPS works is similar to other content protection schemes, in the sense that it uses different classes of encryption key, one type stored on the media (in a **Disc Key Block**) with matching keys stored in the player for the media (which might be based on hardware or software).

Devices that use VCPS have a **Device Key** stored in their internal software (firmware), and use a 128-bit AES cipher to encrypt and decrypt video in the form of MPEG-2 data blocks that are stored on the disk. DVD+RW disks take advantage of a pre-existing system called **Address in Pregroove** (ADIP) that is normally used to calibrate the motor speed of the DVD player by measuring track wobble. This ADIP data is encoded in the disk during manufacturing. VCPS-capable disks have supplementary information, called the Disk Key Block, buried in the ADIP section of the disk. In combination with a **hardware device key** stored in a VCPS device, the Disk Key Block allows an encryption key to be computed. This is then used as a key for the AES-128 algorithm to decode the content.

Advanced Access Content System (AACS)

The **Advanced Access Content System** (AACS) is a copy protection technology administered by the AACS Licensing Administrator on behalf of a large number of companies in the media creation and distribution markets. The AACS protocol is based on the government standard AES 128-bit encryption algorithm that used for the encryption of content on **Blu-Ray** disks. The algorithm uses a set of secret keys associated with the player device, the program to be decrypted and the hardware medium used to store the copy. These are called the **Device Key**, the **Title Key**, and the **Volume ID**. The Volume ID is stored on the physical storage medium (the disk), but is not readable by external software, only the firmware within the device.

An important aspect of AACS is that different devices can have their own hardware keys stored within the device. If these keys are discovered by unauthorized agents and publicized, they can be revoked (via a list of bad keys on future media) so that titles released in the future will not work with them. This has happened to several models of software-based DVD player which were partially decompiled and used to create unauthorized, but effective, open source player software. Unlike CSS, AACS uses Device Keys that can be different for each individual player device, as opposed to each model, so that a single player's key can be revoked.

AACS includes provisions for "end to end" data protection extending from the recording medium to the display device. In order to prevent duplication and recording of content using analog video channels, AACS protocol supports a data flag called the **Image Constraint Token** (ICT) that can be set for a particular program, just as with HDCP. If the ICT is set, any AACS licensed player must not provide the content in analog form at a resolution of better than 960x540, and only provide full HD resolution via HDMI outputs which are protected with HDCP (described above). AACS in later revisions also includes provisions for **Managed Copies**, in which a recording device connects to an Internet site to request permission to make a duplicate of protected content. A copy can be made only if the site grants permission.

The overall security of AACS remains a topic of some debate. As noted above, player keys have been extracted and used to create unauthorized software that could play a limited number of titles. A company called Slysoft was the first to release commercial player software that could play all available titles at the time of its release and apparently remove all encryption from content on Blu-Ray disks.

Video Encoded Invisible Light

VEIL (video encoded invisible light) is not technically a copy protection scheme, but rather a system for placing invisible tracking data into video data. Systems that hide identification information within the image or video content are called watermarking systems, by analogy with the embossed almost-invisible marking on high-quality bond paper. VEIL could be used to place identifying information on every recording made by consumer electronics devices, allowing the source of unlicensed recording to be determined (at least in principle).

There have been proposals to make VEIL watermarking mandatory for recording devices in the United States. The details of how VEIL operates are not public, but it appears that the watermark may lead to detectible effects when it is used. Despite this, VEIL has been identified as a technology to help plug the analog hole by making analog copies harder to produce.

VEIL is also being promoted as a system to allow video data to include "callouts" to the Internet or other devices. For example, a person viewing a video from DVD could take a picture of a video using a digital camera, and this camera could, in principle, extract an embedded VEIL watermark, transmit it to a computer, and access a matching web page.

12. PVR Receivers

Receivers for television broadcasting come in many guises. In this chapter we will touch on a few of the most generic and the most technically notable. The Personal Video Recorder (also known as a Digital Video Recorder or **DVR**) is a device that records video content on a hard disk and typically offers features like elegant on-screen menus of recordings, or simultaneous playback while recording new content. PVR devices are becoming increasing popular, and tend to profoundly change the television viewing experience. The first companies to produce commercial PVR devices were **Tivo** and the now-defunct **ReplayTV**. Stand-alone commercial "media appliances" such as those made by Tivo are the most powerful types of PVR, but there are a variety of related solutions. Since digital receivers are essentially computers limited only by software and disk space, and hard disk prices have dropped substantially, many receivers are available with integrated PVR functionality.

Computer-based applications

A variety of Personal Video Recorder (PVR) software systems are available to control and deliver television and video using a standard home computer. The services that such software delivers can be divided into a few main features directly related to video content, plus all kinds of extra bells and whistles such a providing an alarm clock, photo viewer, email notifications, or access to video games. The key video-related services are:

- Control of the receiver (tuner, satellite receiver, etc.), including channel selection
- Decryption and conditional access support

- Video recording
- Playback and management of recorded content
- Display of an electronic program guide
- Timed recordings

There are PVR software packages for all major operating systems, and there are both open source (free) solutions, as well as commercial packages on the market. Some of the main packages are discussed below, but there is a constant ebb and flow in the available software.

Many computer-based PVR solutions can work in conjunction with a set top box or independent receiver. In order to make recordings, the computer needs to be able to switch channels on the receiver, and this is usually accomplished with an **IR Blaster**. This is a light-emitting diode that is plugged into to the computer/PVR and which beams light into the remote control input of the receiver being controlled, so that the computer/PVR can control the receiver in the same way that a remote control would.

Windows media center

Windows Media Center is an application made by Microsoft Inc. for its Windows operating system. It was first delivered as an add-on for Windows XP, and subsequently bundled with Windows Vista. It provides an interface to a large range of visual media spanning stored photographs, audio recording and live video. As a commercial product integrated with the operating system, it provides a clean well-supported experience with an aesthetically attractive graphics interface. It supports HDTV content for US users only. It also includes a web-based program guide (WebGuide), and supports timer-based recording of shows. An innovative aspect of WebGuide is that it automatically supports the ability to access content stored on the computer using a built in web server.

As a proprietary commercial application, Windows Media Center cannot be as readily modified by the home user as some of the open source solutions discussed below. On the other hand, it does support a powerful "plugin" architecture that allows extra features to be added. Plugins have been used to add support for weather information pages and to allow the viewing of web-based content. In addition, the application itself is highly polished and is well integrated with the operating system and the functions it can provide.

As a licensed commercial application, Windows Media Center respects various types of digital rights management rules, including some which may impede a user's ability to manipulate content in ways they are legally entitled. In some cases, it has been necessary to purchase additional software to allow Windows Media Center to display commercially purchased DVD content, since it must use proprietary CSS decryption for this material (this may only apply to versions for Windows XP). In other cases, playing back DVD content stored on hard disk may involve modifications to the Windows registry.

SageTV

SageTV is a commercial software package that turns a home computer into a DVR allowing it to watch and record live television, play videos, and play back stored recordings from hard disk or DVD. It is generally considered by its users to deliver a high quality user interface. While SageTV was originally written for Windows, it also works on Mac OS X and Linux. This multi-platform support is almost unique among PVR programs.

SageTV supports a smaller set of hardware configurations than many of the competing PVR packages, partly because it depends on cards that include hardware-based MPEG decoding (such as those made by Hauppauge).

BeyondTV

BeyondTV is a commercial computer-based PVR package from SnapStream Media that runs on Microsoft Windows and allows a desktop computer to record video. It provides standard PVR features such as recording, playback, and the ability to pause live television. It supports a good selection of standard video cards for both standard definition and

HDTV. It supports the use of programming via over-the-air broadcasting as well as video input from set-top boxes controlled by an IR blaster.

Some of the exotic features of BeyondTV include semi-automated commercial skipping using automatically-inserted chapter markers, DVD burning of content, web-based administration and scheduling, and the easy transfer of material to portable player devices.

VDR

The **VDR** program is an open source digital video player originally developed for use with satellite broadcasts and the DVB-S protocol. It was developed by Klaus Schmidinger. Over several years of development its capabilities have grown, partly due to an extensible plugin architecture that encourages 3rd-party development by a substantial community of developers. As a result, VDR now supports a variety of other features such as DVD and Mp3 playback, although its primary function is still satellite broadcast reception.

The VDR system design is quite flexible and it can be used with a wide selection of satellite providers, dish types, and hardware devices. While it has many PVR-like features, it focuses on satellite-based programming and makes for a lightweight self-contained package. In general, hardware support is provided via the **Linuxtv** driver framework (described below, see page 203), and the program will work with whatever the hardware driver supports, including a wide range of DVB cards. The program works readily with unencrypted programming. Access to encrypted programming is also possible for devices that support an external conditional access module (CI-interface) smartcard, making it particularly popular in Europe where it can be used with commercial subscriptions. There are also 3rd party software extensions that allow for signal decryption even without a smart card, but they are of dubious legality even if a subscription is held, and they are generally offered as part of an underground software plugin called a **Softcam** (or "sc").

Like many Linux solutions, VDR can be relatively complicated to install and configure. On the other hand, there are also complete ready-to-run installations that combine VDR and a pre-configured operating system to make this straightforward.

MediaPortal

MediaPortal is an open source PVR package for use on Microsoft Windows. It supports numerous types of television reception including DVB-T, DVB-C, DVB-S, ATSC as well as analog TV cards. Much of this functionality and device support is based on the support provided by the Windows operating system.

MediaPortal records files using DVR-MS file format, which is a wrapper for standard MPEG data that can then be exported to other types of software. MediaPortal supports alternative skins (visual appearances), management of stored recordings, photographs, and bookmarks, and access to other programs on the computer being used.

GB-PVR

GB-PVR is a software-based PVR package for Windows. It is free, but not open source. One notable feature, in comparison with other PVR packages for Windows, is that it appears to run quite well on systems with only limited resources (such as Pentium III systems). It supports DVB-S, DVB-T, DVB-C and ATSC and a broad selection of capture cards. The television interface is very complete, but the system includes fewer non-TV functions such as game emulators, than other open source PVR projects.

As with other Windows-based PVR applications, the actual playback is performed using the Microsoft **Directshow** programming interface to the operating system's video streaming architecture. Only analog capture cards with hardware-based MPEG encoders are supported.

MythTV

MythTV is probably the most well established free computer-based television application, integrating a wide selection of home entertainment applications ranging from a DVD player to computer game emulators. It has been under development since 2002, when the project was started by Isaac Richards. It is an open source application that runs on Linux and it provides all the services of a commercial digital video recorder including playback of stored recordings and the display of electronic program guides for live broadcasts. It allows a user to pause live TV, play video games, or record one program while watching another.   Some of the functions of a MythTV system are achieved through the use of other open source tools and libraries besides the Linux kernel itself.

MythTV divides the processing of providing television and home entertainment into **front end** and **back end** services. The front end is responsible for running the graphical user interface and the display of images on screen, while the back end does recording, TV reception, disk storage and management of guide information. The front end and back end services can be split between different computers, if desired, so that the back end computer can be big, complicated and potentially noisy, while the front end computer can be minimized.   This also allows several front end clients to be served by a single back end.   In practice, many users just run both front and back end services on a single standard computer.   MythTV can be configured to work with a wide variety of hardware cards for television reception and display.   This includes cards for terrestrial television, cable TV, and satellite television.   It provides a graphical on-screen guide of upcoming televisions shows and recorded content, and allows for simultaneous playback and recording if the hardware being used supports it.

MythTV stores recorded video in one of two different formats, depending on whether a hardware-based MPEG encoder is in use.   Software encoded files are stored in the rather esoteric **NuppelVideo** file format. The files store MPEG-4 data and can be played by a few other open source applications, but this format is otherwise used by almost nothing else.   Files encoded (or received) as MPEG-2 by hardware cards are stored as raw MPEG-2 data streams.

In general, MythTV requires the use of the **X Windows** system to display graphics on the computer monitor (as well as the TV output), and it uses a **MySQL** database server to store and manipulate the program guide. The raw data for the guide can be obtained from various sources, depending on what country the information is for. Users in the USA and Canada most commonly obtain multi-week program guide information from a subscription service called "**Schedules Direct**" for a cost of about $20/year. In Europe, the free "**nexTView**" service is most commonly used, and **XMLTV**-formatted data services are available for many other regions (XMLTV refers to both an XML-based data format for TV listings as well as the set of tools for manipulating them).

Configuring MythTV from scratch requires a substantial level of technical skill, but there are pre-packaged all-in-one CD installations available that make using it very easy. They automate many of the installation choices and attempt to automatically identify and configure the hardware. Some of the best established such distributions are **KnoppMyth**, **MythDora** and **MythBuntu**, based respectively on the Linux distributions Knoppix, Fedora Core and Ubuntu. Due to the rich feature set, and the use of X Windows, MythTV can require more computing resources that some open source alternatives.

Freevo

Freevo is a Linux-based open source system that uses a PC as a complete home media-management environment. It includes the management of photographs and digital audio, as well as television and stored video. It supports all forms of DVB television (DVT-T/C/S), assuming the suitable hardware is in place, and that the device is supported by the Linux DVB driver. Like open source media players, Freevo achieves much of its power by leaning very heavily on other open source tools within a framework and menu system that it provides.

Freevo is largely written in the Python programming language. This can make it easier to modify than competing systems for those familiar with programming in that language, but it also exacts a penalty in performance.

On the other hand, Freevo makes slightly fewer resource demands than its closest competitor MythTV. Unlike MythTV, it can be operated without the use of the X Windows windowing system or the MySQL database server (using the less resource intensive SQLite library instead). It probably does not scale up as well to have exotic applications, such as those with multiple clients. Freevo was developed for Linux and works well on Mac OS X. It can also be used in a limited form under Microsoft Windows. Installation of Freevo is conditional on a large set of related open source tools, but they can be installed automatically on Linux and Mac OS X.

Windows DVB applications

There is a substantial ecosystem of programs running under Microsoft Windows that provide media acquisition with television tuner control. These programs vary in status from being open source to commercial products, and their user communities vary in scale from small to moderately large. In addition, some of the applications focus on generic media management, some focus on recording of broadcast content, and some may even encourage piracy of intellectual property.

A few of the most popular programs from the many offerings are the **MyTheatre**, **ProgDVB** and **RitzDVB**.

MyTheatre is a commercial Windows-based program from Saar Software that is primarily aimed at satellite-based DVB viewing and recording. The software is commercial, but it also seems to be popular with those users seeking to circumvent conditional access requirements without purchasing a subscription (i.e. signal pirates). Perhaps as a result of this user community, both the company status and the legal status of the package are difficult to determine.

MediaPortal is a full-featured open source PVR application. ProgDVB and RitzDVB both offer a range of features for the control of DVB programming, and have a general style that makes them better suited to the technology aficionado and hacker than for the typical home user. **ChrisTV PVR** provides a large set of PVR features especially for analog video, but also including MPEG content. CyberLink **PowerCinema** is a commercial application that provides full PVR functions, and is available with an included tuner card to assure compatibility.

General-purpose FTA receivers

A range of companies make **Free-to-air** (**FTA**) satellite receivers, meaning receivers that can be used to watch unencrypted content transmitted using the DVB standard protocol. In some cases, these devices may also support other types of data source, such as the reception of terrestrial television. Many of these receivers include standard DVB-CI interface modules, so that subscribers can use them even for encrypted content by inserting an authorized access card.

It cannot be overlooked that many of these FTA receivers can also be modified using unofficial firmware changes that enable them to receive encrypted content, even without a legitimate subscription. Such modified firmware usually only works for a limited amount of time, must be re-modified on a regular basis, and can (in principle) lead to unrecoverable problems. In the USA, in particular, it is probably illegal to sell, or perhaps even to use, this kind of modified firmware. Despite that fact, commercial enterprises that sell receivers that include such firmware pre-installed seem to appear on a regular basis (although how well the devices work is impossible to verify).

It appears that the market for such devices is in constant flux, and is occupied by a range of both a range of conventional legitimate companies such as Philips, CaptiveWorks, Pansat, Viewsat, Sonicview and Patec, as well as occasional small "fly by night" vendors. There are ongoing suggestions that "black market" **clone receivers** are also sometimes sold, those being unofficial unsanctioned copies of some other company's device. The market for FTA receivers in Europe, where they can be used with legal paid subscriptions, seems especially healthy. Many include full PVR functionality, and support HD content as well as standard definition. The range of receivers available includes some with very substantial full-blown computers running the Linux operating system.

Media appliances

While video tape recorders have been in use for many years, the consumer digital video recorder as a self-contained unit is a fairly recent invention; it had to await the development of several fairly recent technologies (such as huge affordable disk drives). Two companies share the credit for developing the first self-contained PVRs: Tivo and ReplayTV. ReplayTV unfortunately stopped producing set-top devices, but they are sufficiently significant to warrant discussion. As PVR technologies drop in price, we can expect to see them integrated into more and more devices, such as television set-top boxes, televisions themselves, and, eventually, home appliances of other kinds. By and large, self-contained PVR appliances offer only a few advantages over computer-based systems, but these few advantages are significant.

Self-contained units tend to have better engineered software systems than many software packages for PCs. There are several explanations for this, but the most compelling are that the long-term resources and incentives to develop good software are better, and that the close integration of the hardware and software allows for a better overall experience. Self-contained PVR units also tend to have a few exotic features, such as high quality electronic program guides or automated systems that provide recommendations to the viewers, and as part of a developing trend, the ability to download or purchase content on demand.

Tivo

Tivo is the brand name that sets the standard for PVR technology in terms of history, functionality and brand recognition. Their various PVR offerings are generally regarded as the most elegant and effective examples of the technology, and they have been licensed to other companies and integrated into other products, such as cable TV decoders. Tivo units require a monthly paid subscription to operate. Among the most notable and specialized feature of Tivo devices is the ability of the system to automatically determine, suggest and record content that infers may be interesting to a viewer based on their past viewing habits. Such

recommender systems are based on the analysis of the viewing habits of many different subscribers.

Shortly after Tivo products were introduced a substantial community of Tivo hackers developed. These users added many features to their devices, primarily the **Series 1** Tivo models. This activity has been curtailed over the last few years due to the greater use of encryption in more recent Tivo models and firmware features that prevent the modification of the unit. Recent Tivo units, for example, require de-soldering and modification of a of a PROM chip before custom software can be installed by the owner.

ReplayTV

The **ReplayTV** device was one of the first PVR systems, but is no longer made as a stand-alone unit. The system introduced some innovative features such as Internet streaming video and the fully automated deletion of commercials. These features led to legal disputes, which may be partially responsible for the financial difficulties encountered by the company. ReplayTV devices, like older-model Tivo units, store their data on an unencrypted file system, which means that recording can be recovered. Furthermore, ReplayTV units support Internet-based protocols (HTTP) that allow content to be transferred to another device using free software tools. Particularly as a result of these unusual features, ReplayTV devices are still traded on the second hand market, and are sometimes touted as the best of their kind.

Slingbox

The Slingbox exemplifies the network hub whose primary strength is the distribution of recorded or live media across a local network or the Internet. It is not so much a PVR, but a device that provides a device-agnostic interface between a PVR or other video source and the Internet at large.

DTV operating system drivers

Windows

Microsoft Windows contains an application programmer interface (API) and related libraries and operating support system called DirectShow. This provides an API to operating system support for streaming digital video, including support for DVB, ATSC, and analog video over satellite, digital cable and terrestrial broadcast. While at one time the development tools for this interface were part of the **DirectX** software development kit (SDK), they have migrated into the Microsoft Windows SDK proper. Software to be used with this interface must be designed to be compatible with the Microsoft **Broadcast Driver Architecture** (BDA).

One part of DirectShow is the Microsoft **TV Technologies Application Interface**. This includes an interface for creating video controllers using ActiveX, an API for tuning in stations that works across different hardware devices, a generic API for video input devices, and a system for presenting video on output devices. It includes, for example, an MPEG-2 "Transport Information Filter" that allows selected packet types, based on PID values, to be extracted from the data stream and passed on to appropriate handlers. The majority, if not all, of the Windows-Based applications for streaming video make use of this API for video input at least. On the other hand, it appears that a large fraction of existing DVB and ATSC applications contain custom code to parse the EIT, PMT and other tables, perhaps because these application were deployed while official support for such functionality was not yet available.

The Broadcast Driver Architecture is the critical component that links the hardware to the operating system. Hardware requirements for this API include a BDA-compatible digital TV tuner, a DirectX Video Acceleration-compatible video card with 32 MB VRAM, and a DirectX VA-compatible MPEG-2 decoder.

In order to provide digital rights management and conditional access control, the **Protected Broadcast Driver Architecture** (PBDA) was developed. It exists in two forms, the **PDBA-KS** for content control within a single PC, and the **PDBA-IP** for content control based on streaming data from external sources such as cable or satellite. The PDBA architecture is meant to support a large suite of content control standards for digital rights management. These include CGMS-A, MacroVision technologies and the Broadcast Flag, and protection of locally stored content with encryption based on Windows Media Digital Rights Management (**WMDRM**). The PDBA system recognizes incoming content and stores it using WMDRM, including flags indicating how it can be manipulated (i.e. whether copies permitted). WMDRM combines AES-128 cryptography, HDCP transmission control, and uses the RSA algorithms for the exchange of cryptographic keys with other devices. WMDRM content is kept encrypted on the computer and is only displayed digitally using a secure (encrypted) mechanism such as HDCP.

Mac OS X

Macintosh software for digital television is primarily dependent on USB devices using custom USB drivers, which in turn interoperate with Apple's **QuickTime** video architecture.

Linux

Linux drivers for DVB and ATSC are part of the open source **Linuxtv** project. Digital television devices are supported via the **dvb-kernel** module, which includes support for DVB-S, DVB-T, DVB-C, and ATSC. Most applications also take advantage of the closely related **Video4Linux** (v4l) kernel system to provide real time video processing and support for additional hardware cards. This kernel module provides facilities for extracting packets based on PID value, processing video data and other standard activities.

The developer community for this software seems fairly active, and so a wide range of hardware devices is supported both via the PCI bus and USB. As is often the case, however, much of this functionality is provided without support from the hardware manufacturers. Thus it depends on reverse engineering by the open source community and may not be available for the latest hardware. That issue aside, the suite of Linux driver features for many hardware platforms meets or exceeds what is available on commercial operating systems.

13. Digital TV via home computer

Digital television is based on data packets that contain video information. What this means is that the hardware devices that support digital television are basically specialized computers, or specialized computer networking appliances.   A key side effect of this is that most ordinary desktop computers can make adequate, and typically terrific, home media centers or television receivers.

The one atypical component that a computer needs to receive and process digital television is a tuner/receiver card to capture the radio frequency signals that transmit the television programming (unless all you want to manipulate is internet-based streaming video or recorded content). This means inserting either an internal card into the computer to receive content, or else plugging in an external receiver box via either the USB connector or, much more rarely, a Firewire connector.

The computer then needs to run suitable software to control the receiver and to manipulate the content. The best and most commonplace software plays the role of a Personal Video Recorder (PVR) which not only allows channels to be selected, but also includes many other great features.

Hardware receiver cards

Computer-based support of digital television is a fairly recent innovation. While analog satellite broadcasts such as those on C-band can be digitized and displayed with any video capture card, the computer only acts as a display and video capture device and all the satellite-specific processing is accomplished by a proprietary set-top box. Even some modern satellite-based systems like Videocypher II are based on proprietary solutions and analog signal processing, and so essentially the only solutions that are available are those delivered in sealed set-top boxes that take a transmitted signal as input, and deliver a (single) video program out. As a result, we will restrict our attention in this chapter to newer, non-proprietary digital transmission protocols.

Even with standard protocols such as DVB and ATSC, a large fraction of the available programming is encrypted in proprietary ways. Fortunately, the protocols are rich enough to support standard data delivery with encrypted data wrapping inside standard data packets. As a result, commercial solutions are available for handling a mixture of both unencrypted free-to-air programming that can be readily displayed, encrypted data that the box may be able to handle under certain circumstances, as well as encrypted data that cannot be viewed. In addition, some users may find a broadcaster's metadata (such as the EPG) of interest, even when the programming associated with it is encrypted.

Many hardware devices for use with computers support a special slot for holding a manufacturer-specific decryption card that comes with your programming subscription, and thus can handle encrypted signals (if you subscribe and pay for them) as well as unencrypted content. For satellite programming, this support for subscription-based channels is almost invariably provided via a smart card and a suitable interface slot. To receive cable television, a QAM tuner card is required for unencrypted programming. Only a few solutions for encrypted programming are available, primarily those based on the use of a CableCARD. In order for a desktop computer to support a CableCARD though, the entire computer needs to be certified by Cable labs; simply inserting a digital cable tuner (called an OCUR) into an ordinary computer, even with an approved

CableCARD, will not work. CableCARD certified computers are generally more costly, and include specialized firmware in the BIOS as well as other data-protection features.

For unencrypted signals, however, all that is needed is a stand-alone tuner card that can be inserted into almost any standard desktop PC (i.e. a computer with an Intel/AMD-type internal architecture). Furthermore, several broadcasters support encrypted subscription-based programming delivered via satellite with the use of a Conditional Access Module that comes as part of many tuner cards (see page **155**). North American broadcasters, however, have opted for proprietary encryption technologies so that they cannot generally be used with standards-based receivers, even using a paid subscription, unless substantial unlicensed modifications are made to the standard software. This means that commercially available solutions for computer-based reception in North America are confined to either plain video capture (for example using set-top boxes with analog output), free-to-air unencrypted programming, or European programming using a subscription to a European broadcaster (which is generally cannot be obtained in much of North America). There is also a steady market in pirate decryption devices that subvert commercial restrictions on programming, but the sale of such devices is never sanctioned by the broadcasters and is in a legal grey zone, or even illegal, in many places most notably the USA. Therefore, since that is all that is commercially available we will focus on over-the-air ATSC, and DVB transmissions for the rest of this chapter.

Computer-based DVB decoders for every possible video format are available as either PCI cards (including PCI express), that go inside the computer chassis, or as USB or Firewire (IEEE 1394 devices for use with standard computers). PCI cards plug into the backplane of the computer just like a video card and often include an output connector for directly driving a regular television set. Cards that support the handling of HDTV content often use PCI-express (PCIe) since the HDTV programming sometimes needs the higher data rates that only the "express" bus can provide. USB devices for television reception usually provide the data to the PC via the USB protocol (using USB version 1.1 or 2.0), but rarely include outputs for directly driving a television set. For USB devices that support high quality HDTV signals, the USB 2.0 protocol is definitely needed.

All classes of device, PCI, PCIe, USB, or Firewire, provide rich sets of features and can be used with Microsoft Windows, Linux and Mac OS X. In general most manufacturers support Windows. Linux support is usually available only on a do-it-yourself basis, but the required drivers are widely available and are built into to several standard kernel distributions. Mac OS support is less common, but does exist (for example from Elgato systems).

In general, USB devices are simple to use when they are supported, and can provide a simple out-of-the-box solution. Manufacturers' supported solutions always come with TV viewing software that does channel selection, and it typically provides delayed recording features to record a show while you are away from the machine. USB solutions are typically closed boxes, both physically and metaphorically, and thus don't provide as great a range of features as PCI systems either off-the-shelf or for those who want to customize the system or use non-standard software. PCI cards, on the other hand, allow the operating system to interact with many of the interface's low-level features, and thus can sometime provide higher data rates and a wider range of exotic features. This can include the ability to record or watch multiple channels from a single satellite at the same time (assuming both channels are on the same transponder). Most PCI-based solutions use one of a fairly small set of basic components to implement the key reception, display, and video decoding functions and as a result there is a substantial degree of compatibility between different hardware cards and different kinds of software. Furthermore, once the device driver for a PCI card is installed, it usually presents a fairly standard interface to the operating system so the viewing software doesn't have to be too highly tailored to the specific hardware card.

Almost all DVB cards with a PCI interface have the same main hardware functions, with a few of them being optional:

- A **front end** module (where front refers to being closer to the antenna) made up of a tuner/receiver,

- A **demuxing** module that separates the incoming transport stream into audio and video packet streams,

- A **conditional access** (CI) **module**, which may permit connection to physically separate conditional access cards that deal with encryption,

- An MPEG **decoder** (optionally) that decodes and decompresses the MPEG-encoded video and audio data,

- A video driver (optionally) that provides a signal that can be connected directly to a television.

The video driver, when present, often includes special features to support an on-screen overlay. This is especially common on cards that have a TV output. Such an on-screen data display (OSD) the works as an overlay on the live video stream allowing for channel data and other information to be shown in real time over the picture.

Some cards omit the video driver and leave it up to the host computer to display the video on the monitor or send it elsewhere. Similarly, cards without MPEG decoders leave it up to the CPU of the host computer to do the MPEG decoding and thus have a lower price tag. As we saw earlier, the data transmitted via the DVB protocol is compressed and encoded using the MPEG protocol, and needs to be decompressed in order to be displayed. This is a substantial computing task, and uses a fair bit of a host system's resources; as a bare minimum it needs about the CPU processing power equivalent to a fully loaded Pentium III 700MHz, with some variation due to code quality and various other factors.

A system with all the options, including hardware MPEG decoding on the receiver card, can act as an almost complete cable-OTA-satellite receiver-decoder while placing almost no CPU demands at all while running on the host computer. Such cards are commonly known as "Premium" cards, or "**FF cards**" (for **Full-featured** although the sole distinguishing feature is the MPEG decoder). Cards without an on-board MPEG decoder are commonly known as **budget cards**. Full-featured cards were clearly the preferred option at one time and still have no disadvantages compared to budget cards, but they have declined somewhat in popularity as host computers have increased in speed. The most common chip used for MPEG decoding in such cards has been the Texas Instruments AV7110 series digital signal processor (DSP), and it is used in cards from several popular DVB card manufacturers such as Hauppauge (and the company they acquired called Technotrend). The TMS320AV7100, for example, is a 240-pin chip that includes a MPEG-2 decoder, several megabytes of SDRAM and an ARM7T 32-bit RISC CPU. This particular chip, while still quite popular, does not support high definition video, though. Thus, even "full featured" cards with this chip (and others like it) need to decompress HD video content using the host CPU. Since Hauppauge and Technotrend have used such a common chip set in many cards, these cards define a *de facto* standard and are sometimes referred to simply as **TT cards** (for Technotrend), either full-featured or budget variations.

The use of a computer equipped with a receiver card, either for DVT-S/T/C or ATSC depends critically on software support. On Windows the support for TV cards is often supplied by operating system drivers, but not always. On Linux, DVB support (including ATSC) is generally supplied via the DVB driver that can be configured as part of the Linux kernel (along with the Video for Linux V4L drivers). With Mac OS, hardware support needs to be installed as part of the kernel, but the required drivers are generally provided along with specific applications or hardware and the most prevalent hardware solutions are based on USB devices.

Since software support is so critical and complex newer manufacturers often try to assure support by providing the same interface (API) as an existing (supported) card. This is especially true for PCI cards where the interface options can be overwhelming otherwise. Many cards behave like the Technotrend cards discussed above. These include (in addition to Technotrend TT cards themselves) the following: TechniSat SkyStar1, Hauppauge WinTV-DVB-S, Hauppauge WinTV-Nexus-S, Fujitsu Siemens PCI DVB Sat, Galaxis plugin-s, the Galaxis DVB card S CI. Another group of cards attempt to achieve compatibility with the TwinHan series of cards, and these include (in addition to TwinHan cards themselves) the Hercules Smart TV Satellite, Prolink PixelView DTV2000, Prolink PixelView DTV3000, PowerColor DSTV, ProVideo PV-911 and the ProVideo PV-911CI.

Most computer-based applications for television support are packed as part of a PVR-like package. These are discussed in the section dealing with PVRs.

14. Sample Satellite Services

Since satellite signals are rebroadcast from space, the satellite programs that can be received in a particular region may not be coming from nearby broadcasters. This is one of the appealing things about satellite television and radio: you can get programming from distant lands. In most countries, however, the selection of subscription-based commercially available satellite programming providers is fairly limited. This limited selection arises due to two factors. Firstly, the cost of owning a satellite, or a portion of a satellite, is very high and that limits the set of companies with the resources to compete. Secondly, many countries have legislated restrictions on who can sell programming, even if it arrives via space-borne satellites. These factors together imply that only a few subscription-based providers usually exist per country, and subscriptions to other companies in other countries many not be permitted.

Even in countries such as Canada where satellite or terrestrial programming from other countries is accessible (in that case from the United States), there are often legal barriers to subscribing to that programming. In the particular case of Canada, the signals from US satellite broadcasters are readily available to most Canadians (since most Canadians live near the US border), but Canadian broadcast regulations do not permit US broadcasters to sell the required subscriptions in Canada. As a result, Canadians are not able to subscribe to US satellite programming unless they do it using a US address provided by a friend or relative in the USA. The same applies to the purchase of Canadian programming by US customers. This process of subscribing via a third-party address is referred to as a **grey market** subscription, as it involves cross-border sales of programming (or hardware) that is probably not technically illegal, yet neither is it sanctioned by the provider.

In addition to subscription-based satellite programming, a large quantity of satellite-based programming is available without any subscription, just like unencrypted terrestrial broadcasting. This unencrypted programming is called Free-To-Air (FTA) and exists on both the Ku band and C band. Free-to-air programming includes the C-band wild feeds discussed earlier, but is more commonly used to refer to Ku-band broadcasts as well as to a small number of unencrypted channels that may be available on DBS satellites. By and large, FTA channels are less consistently reliable than commercial DBS programming since they include special-purpose programming and all kinds of non-standard material. On the other hand, FTA programming includes broadcasts in a wide selection of languages. No matter what country you are in, FTA programming provides material from other countries. In general, North America has access to less FTA programming than most other continents, but North American viewers can still receive programming from almost a dozen satellites that carry hundreds of FTA channels in languages ranging from Arabic to Vietnamese.

Some of the reasons that viewers have cited for watching FTA programming include the pleasure of tinkering with the technology, the access to foreign language or cultural programming, the access or programs not available in a local market, the access to a diverse selection of news programming, and the availability of specialized niche programming.

The table below provides an example of some of the channels in diverse languages that are available with FTA programming in one particular month (this table puts particular emphasis on North American Ku-band coverage).

Channel or Broadcaster Name	Language (listed language may be just one of several options)	Satellite name
Ethiopian TV	Amharic	Galaxy 25
2M TV	Arabic	Galaxy 25
Abu Dhabi Sports	Arabic	Galaxy 25
Aghapy TV	Arabic	Galaxy 26
Al Alam News	Arabic	Galaxy 25
Al Anwar	Arabic	Galaxy 25
Al Fayhaa	Arabic	Galaxy 25
Al-Forat	Arabic	Galaxy 25
Al-Iraqiya	Arabic	Galaxy 25
Alkarma TV	Arabic	Galaxy 25
Almaghribya TV	Arabic	Galaxy 25
Aria International TV	Arabic	Galaxy 25
Canal Algerie	Arabic	Galaxy 25
Commercial TV	Arabic	Galaxy 25
Jordan TV	Arabic	Galaxy 25
Kuwait TV	Arabic	Galaxy 25
Libyan Jamahiriya Broadcasting	Arabic	Galaxy 25
Middle Eastern Broadcasting Net	Arabic	Galaxy 25
Orange TV	Arabic	Galaxy 25
Payame Afghan TV	Arabic	Galaxy 25
Qatar TV	Arabic	Galaxy 25
Salaam TV	Arabic	Galaxy 25
Saudi Arabian TV 1	Arabic	Galaxy 25
Sharjah TV	Arabic	Galaxy 25
Sudanese Space Channel	Arabic	Galaxy 25
Sultanate of Oman TV	Arabic	Galaxy 25
Syria Satellite	Arabic	Galaxy 25

Channel or Broadcaster Name	Language (listed language may be just one of several options)	Satellite name
Channel		
Toronto Entertainment Network	Arabic	Galaxy 25
Tunis 7	Arabic	Galaxy 25
Yemen TV	Arabic	Galaxy 25
Aramaic Broadcasting Network	Aramaic	Galaxy 25
AssyriaSat	Aramaic	Galaxy 25
Ishtar TV	Aramaic	Galaxy 25
MRTV/ABN	Aramaic	Galaxy 25
Suroyo TV	Aramaic	Galaxy 25
Armenian/Russian TV Network	Armenian	Galaxy 25
Horizon Armenian TV	Armenian	Galaxy 25
Azerbaycan Televiziya	Azerbaijani	Galaxy 25
CCTV 4	Chinese	Galaxy 3C
Da-Ai TV	Chinese	SatMex 6
Da-Ai TV 2	Chinese	Galaxy 25
Hwazan Satellite TV	Chinese	SatMex 6
Hwazan Satellite TV	Chinese	Galaxy 25
IAVC	Chinese	SatMex 6
IF TV	Chinese	SatMex 6
New Tang Dynasty TV	Chinese	Galaxy 25
Sun TV	Chinese	SatMex 6
Taiwan Macroview TV	Chinese	Galaxy 3C
Taiwan Macroview TV	Chinese	SatMex 6

Channel or Broadcaster Name	Language (listed language may be just one of several options)	Satellite name
Ariana TV Network	Dari	Galaxy 25
Noor TV	Dari	Galaxy 25
Radio and TV of Afghanistan	Dari	Galaxy 25
BVN-TV	Dutch	AMC 4
3ABN	English	AMC 4
3C Community	English	Galaxy 10R
3CTV	English	Galaxy 10R
ABC	English	L6 (occasional)
ABC News Now	English	Galaxy 28
Al Jazeera English	English	Galaxy 25
Angel One	English	EchoStar 7 (DBS)
Apostolic Oneness Net	English	AMC 4
Apostolic Oneness Network	English	Galaxy 3C
Ariana Afghanistan TV	English	Galaxy 25
AZCAR Training	English	Galaxy 25
BBC America	English	C3 (C-band)
BYU TV	English	Galaxy 28
CBC Canada	English	F1 (C-band)
CCTV 9	English	Galaxy 3C
Church Channel, The	English	Galaxy 25
Create TV	English	AMC 3
Daystar	English	Galaxy 10R
Daystar	English	Galaxy 25
Ebru TV	English	Galaxy 25
Emmanuel TV	English	Galaxy 25
Emmanuel TV (716)	English	Galaxy 25
Eurosport	English (and many other languages)	Sirius 4
Euronews	English (and	Galaxy 23

Channel or Broadcaster Name	Language (listed language may be just one of several options)	Satellite name
	many other languages)	
Fashion TV America	English	Galaxy 25
Federal Judicial TV Net	English	Galaxy 26
Global Christian Network	English	AMC 4
God's Learning Channel	English	Galaxy 25
Gospel Broadcasting Network	English	AMC 6
Gospel Music TV	English	AMC 4
Hope Channel	English	AMC 4
Infomercials	English	Galaxy 25
Infomercials 2	English	Galaxy 25
Infomercials 3	English	Galaxy 25
Infomercials 4	English	Galaxy 25
Infomercials 5	English	Galaxy 25
JCTV	English	Galaxy 25
KATV ABC Little Rock AR	English	Galaxy 10R
KBTZ Fox Butte MT	English	Galaxy 10R
KCBU RTN Salt Lake City UT	English	Galaxy 10R
KDEV-LP RTN Aurora CO	English	Galaxy 10R
KEGS RTN Las Vegas NV	English	Galaxy 10R
KFDF RTN Fort Smith AR	English	Galaxy 10R
KFTL ind Oakland CA	English	AMC 6

Channel or Broadcaster Name	Language (listed language may be just one of several options)	Satellite name
Kingdom of Jesus Christ	English	Galaxy 25
KKTU ABC Cheyenne WY	English	Galaxy 10R
KKYK RTN Camden AR	English	Galaxy 10R
KLMN Fox Great Falls MT	English	Galaxy 10R
KMMF Fox Missoula MT	English	Galaxy 10R
KPBI MyNet Fort Smith AR	English	Galaxy 10R
KQUP RTN Spokane WA	English	Galaxy 10R
KTVC RTN Eugene OR	English	Galaxy 10R
KTWO ABC Casper WY	English	Galaxy 10R
KUIL Fox Beaumont TX	English	AMC 4
KWBF MyNet Little Rock AR	English	Galaxy 10R
KWBM MyNet Hollister MO	English	Galaxy 10R
KWCE RTN Alexandria LA	English	Galaxy 10R
KWWF RTN Waterloo IA	English	Galaxy 10R
LLBN	English	AMC 4
LoveWorld	English	Galaxy 25
Maharishi Channel	English	Galaxy 25
MHz Worldview	English	Galaxy 25
Miracle Channel, The	English	AMC 4
Montana PBS	English	AMC 3

Channel or Broadcaster Name	Language (listed language may be just one of several options)	Satellite name
Muslim TV Ahmadiyya Intl	English	AMC 3
NASA TV	English	EchoStar 7 (DBS)
New York Network	English	AMC 5
Nigerian TV Authority	English	Galaxy 25
Ohio News Network	English	SBS 6
Patient Channel, The	English	AMC 3
PBS DTV	English	AMC 3
PBS NPS 1	English	AMC 3
PBS NPS 2	English	AMC 3
PBS NPS 3	English	AMC 3
PBS NPS 4	English	AMC 3
PBS X	English	AMC 3
PBS XD	English	AMC 3
PBS XP	English	AMC 3
Peace TV	English	Galaxy 25
Pentagon Channel	English	AMC 1
Press TV	English	Galaxy 25
Research Channel	English	Galaxy 10R
Russia Today	English	Galaxy 25
Saudi Arabian TV 2	English	Galaxy 25
Smile of a Child	English	Galaxy 25
Spirit Word Channel, The	English	Galaxy 25
Supreme Master TV	English	Galaxy 25
Tip TV	English	AMC 3
Transforming Lives Network	English	Galaxy 10R
Trinity Broadcasting Network	English	Galaxy 25
University Network, The	English	Galaxy 26
University of	English	Galaxy 10R

Channel or Broadcaster Name	Language (listed language may be just one of several options)	Satellite name
Washington TV		
Veterans Affairs Knowledge Network	English	Galaxy 10R
Vision TV	English	Galaxy 25
WABC New York (ABC)	English	K1
WBLU MyNet Lexington KY	English	Galaxy 10R
WGMU MyNet Burlington VT	English	Galaxy 10R
White Springs TV	English	Echo 5/Galaxy 27
WMQF Fox Marquette MI	English	Galaxy 10R
WNGS RTN Buffalo NY	English	Galaxy 10R
Word Network, The	English	Galaxy 25
WOUB PBS Athens OH	English	Echo 5/Galaxy 27
WPXS RTN Mount Vernon IL	English	Galaxy 10R
America Farsi Net	Farsi	Galaxy 25
Andisheh TV	Farsi	Galaxy 25
Appadana International	Farsi	Galaxy 25
Armaghan TV	Farsi	Galaxy 25
Channel One TV	Farsi	Galaxy 25
Didar Global TV	Farsi	Galaxy 25
Iran TV Network	Farsi	Galaxy 25
Iranian News Network	Farsi	Galaxy 25
Jaam-E-Jam International	Farsi	Galaxy 25
Jame-Jam TV	Farsi	Galaxy 25
Khorasan TV	Farsi	Galaxy 25

Channel or Broadcaster Name	Language (listed language may be just one of several options)	Satellite name
Live Channel	Farsi	Galaxy 25
Nejat TV	Farsi	Galaxy 25
Omid-e-Iran	Farsi	Galaxy 25
Pars TV	Farsi	Galaxy 25
Payam TV	Farsi	Galaxy 25
Persian Bazaar TV	Farsi	Galaxy 25
Persian Entertainment Network	Farsi	Galaxy 25
Persian TV	Farsi	Galaxy 25
Rang-A-Rang TV	Farsi	Galaxy 25
Simay-Azadi	Farsi	Galaxy 25
Tamasha International Network	Farsi	Galaxy 25
Tapesh TV Network	Farsi	Galaxy 25
CCTV Français	French	Galaxy 3C
Tele Citronelle Satellite TV	French	Galaxy 25
Duna TV	Hungarian	Galaxy 25
Tele Radio Padre Pio	Italian	Galaxy 25
Telepace Roma	Italian	Galaxy 25
Christian Global Network	Korean	AMC 4
Korea Biz Joint	Korean	Galaxy 25
Kurdistan TV	Kurdish	Galaxy 25
KurdSat	Kurdish	Galaxy 25
Lao Champa TV	Lao	Galaxy 25
Lao TV5	Lao	Galaxy 25
LTV World	Lithuanian	Galaxy 25
Telewizja Trwam	Polish	AMC 4
RTP Internacional America	Portuguese	AMC 4
TV Canção Nova	Portuguese	Galaxy 25

Channel or Broadcaster Name	Language (listed language may be just one of several options)	Satellite name
Internacional		
TV Romania International	Romanian	Galaxy 25
CNL	Russian	Galaxy 25
Impact TV	Russian	Galaxy 25
Rodnoy TV	Russian	Galaxy 25
TBN Russia	Russian	Galaxy 25
3ABN Latino	Spanish	AMC 4
Almavision	Spanish	SatMex 5
Bethel TV	Spanish	AMC 4
CCTV Espanol	Spanish	Galaxy 3C
El Sembrador Nueva Evangelizacion	Spanish	SatMex 5
Edusat	Spanish	SatMex 5
Esperanza TV	Spanish	AMC 4
KAMT TF Amarillo TX	Spanish	Galaxy 10R
KEYU Uni Amarillo TX	Spanish	Galaxy 10R
KMCC Mm Laughlin NV	Spanish	Galaxy 10R
KTEL Tel Carlsbad NM	Spanish	AMC 5
KUTF TF Price UT	Spanish	Galaxy 10R
KUTH Uni Logan UT	Spanish	Galaxy 10R
KUWF Uni Wichita Falls TX	Spanish	Galaxy 10R
KWKO Uni Waco TX	Spanish	Galaxy 10R
KXUN Uni Fort Smith AR	Spanish	Galaxy 10R
Restauracion	Spanish	SatMex 6
TBN enlace	Spanish	Galaxy 25

Channel or Broadcaster Name	Language (listed language may be just one of several options)	Satellite name
Televida Abundante	Spanish	SatMex 5
V-Me	Spanish	AMC 3
WNYI Uni Ithaca NY	Spanish	Galaxy 10R
WUMN Uni Minneapolis MN	Spanish	Galaxy 10R
Dhamma Media Channel	Thai	Galaxy 25
E-san Discovery	Thai	Galaxy 25
Happy Variety Channel	Thai	Galaxy 25
IPtv Music and Shopping	Thai	Galaxy 25
Network of Asian Television	Thai	Galaxy 25
Thai Overseas TV	Thai	Galaxy 25
Thai TV Global Network	Thai	Galaxy 25
E2 Channel	Turkish	Galaxy 25
Markazi TV	Turkish	Galaxy 25
Samanyolu TV	Turkish	Galaxy 25
TRT International	Turkish	Galaxy 25
AAJ TV	Urdu	Galaxy 10R
Hum TV	Urdu	Galaxy 10R
TV One	Urdu	Galaxy 10R
WAQT TV	Urdu	Galaxy 10R
VTV-4	Vietnamese	Galaxy 25

Figure 44: DirecTV's 5-LNB Dish Antenna. Provided by Jeremy Zafran (Creative Commons license).

Selected satellite operators

Although there are numerous broadcasters the world over, there are only a few companies that operate and own the actual satellites that transmit the programming. Free-to-Air broadcasters are commonplace and tend to change on a regular basis, but still provide a key source of non-English broadcasting for North Americans. The Intelsat Galaxy 25 satellite located at 97 degrees West latitude, for example, is a key source of such programming.

Commercial programming is more consistent. An illustrative selection of some of the most popular commercial operators is provided here. In North

America the same companies tend to both provide broadcasting services and operate their own satellites, while elsewhere the programming providers often lease space from a distinct satellite operating company.

DirecTV

DirecTV is one of the two major satellite providers in the United States and also provides programming in several other countries particularly in the Caribbean and South America. DirecTV broadcasting is transmitted using direct-to-home signals which, due to the proprietary systems involved, are referred to as **DSS** (for **direct satellite systems**, see also page 122). They also broadcast outside North America using FSS spectrum, and may be planning a shift towards DVB-S2 in the future. Programming is encrypted and available only by subscription and requires the use of a proprietary receiver and smart card. This broadcaster offers a wide selection of content. Dish network antennas are moving towards a 5-LNB design that allows the receipt of both Ku and Ka signals (from 5 satellites). Their satellites include those at 109.8°W, 72.5°W, 99.2°W, 101.2°W, 102.8°W, and 119.0°W.

DirecTV was a favorite target of US satellite pirates in the late 1990s, being among the first direct-to-home satellite broadcasters, and went through a series of revisions to their smart card technology. The early model called the "F" (or P1) card was among the first to be used, and was eventually thoroughly subverted by a non-commercial "hacker" community that arose on the Internet. Subsequent revisions known as "H-card" and the "HU-card" were also eventually subverted and have become well known as examples of this issue. A new encryption technology was introduced with a smart card model called the "P4" card and continues to be used in the more recent "D1" and "D2" (or "P12") card formats. It appears to be secure from subversion and tampering. DirecTV uses VideoGuard for signal encryption (discussed in greater detail in the context of conditional access systems).

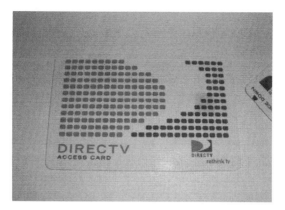

Figure 45: DirecTV D2 access card.

Dish Network

Dish network (owned by Echostar Communications, and sometimes referred to by that name) is also a major US programming provider. Almost all Dish Network programming is encrypted (using Nagravision), with the exception of a small number of public service channels and some music broadcasts. Most programming such as network broadcasts, movies, and pay-per-view films are encrypted. Dish Network uses more-or-less standard DVB protocols for DVB broadcasting. The receipt of subscription programming, while possible in principle with standards-compliant equipment, is only supported using proprietary receivers which, in turn, depend on proprietary smart-card technology. The satellites used include several name **Echostar**, such as EchoStar 8/10 at 110.0°W.

Bell ExpressVu

Bell ExpressVu is the primary satellite programming provider in Canada. It uses DBS broadcasting and employs a selection of satellites, notably at 91°W and 82°W. Programming is encrypted using the Nagravision system and available by subscription only. It includes a wide selection of Canadian content, various specialty channels, and a number of pay-per-view movie channels.

StarChoice

StarChoice provides broadcast services in Canada, although is uses FSS instead of DBS broadcasting.   This is provided via the Anik F1 and F2 satellites originally launched by the Canadian government and leased for commercial use. The unusually small separation between the two satellites mandates the use of a special LNB due to the use of narrowly-spaced satellites so that Starchoice users require a unique dish.   Starchoice programming is encrypted using Motorola's Digicipher 2 (DCII) encryption system and thus the receiver hardware is also incompatible with other broadcasters.

Astra

SES Astra is a major commercial satellite operator that provides multi-lingual DVB satellite programming on the BSS/Ku band to most of Europe. Astra carries a large number of unencrypted subscription-free channels over a dozen satellites, free channels that are encrypted, and encrypted channels that require a paid subscription. All in all, their satellites carry on the order of 1,000 channels. Although Astra signals cover almost all of Europe and beyond, there are legal restrictions on what channels can be subscribed to that depend on national regulations in different countries.

Eutelsat

Eutelsat is a huge European satellite operator that operates numerous satellites (over 20) and leases capacity and services to many other broadcasters. The satellites operate on the Ku band, providing both BSS and FSS transmissions. Several of the satellites operated by Eutelsat have names with the suffix "-bird," such as **Hotbird** and **Eurobird**, and their footprint covers all of Europe, the Middle East, North Africa, and other regions including North and South America. The Hotbird 6 satellite alone (located at 13E), as an example, provides well over 200 unencrypted channels and over 100 encrypted channels.

DStv

Digital Satellite Television, or DStv/Multichoice, is a programming provider primarily serving Africa. Technically, DStv is not a satellite operator *per se*, since the actual satellites that are used for broadcasting are operated by other companies such as Intelsat and Eutelsat. DStv provides content from various national broadcasters spanning the African continent. It uses the Ku band as well as the C band, and provides encrypted and some unencrypted programming.

Arabsat

Arabsat (the Arab Satellite Communications Organization) operates a small fleet of satellites whose coverage extends largely over the Middle East, North Africa, and the Mediterranean basin.

Nilesat

Nilesat is an Egyptian company that operates the satellites of the same name. The satellite Nilesat 102, for example, carries 12 Ku-band transponders and is located at 7°W. Nilesat DBS coverage is aimed primarily at the northern half of Africa and the Middle East. It delivers several hundred channels, with particular emphasis on Egyptian and Arabic content.

AsiaSat

The Asia Satellite Telecommunications Company Limited operates satellites called AsiaSat which provide C-band direct-to-home service covering Asia, Russia, the Middle East and Australasia. AsiaSat uses both the C and Ku bands and reaches over 50 different countries with its transmissions. The programming is provided by some 100 different broadcasters using almost 300 stations and with channels in a wide range of different languages.

BrazilSat

BrazilSat satellites are part of Brazil's national **Sistema Brasileiro de Comunicação por Satélites** (SBTS). They are operated by Embratel and transmit broadcasting in the C-band. Brazilsat B4, for example, is positioned at 92°W and transmits using roughly 28 C-band transponders. The Canadian Space Agency played a key role of the development of BrazilSat 1 and BrazilSat 2.

Intelsat

The International Telecommunications Satellite Organization (INTELSAT) operates on the order of 50 different satellites. The fleet of satellites spans the globe and provides coverage to almost every region using both the C and Ku bands. As examples of the diverse fleet, the satellite Galaxy-3C (G-3C) located at 95°W provides both C-band and Ku-band coverage to North America, the satellite IS-805 at 304.5°E provides coverage in the Ku band to all of the Americas, and Intelsat-602 (IS-602) is in an inclined orbit at 157°E and provides coverage in the C and Ku bands for a range of countries from Australia to Eastern China.

NTV Plus

NTV (written as **НТВ** in Cyrillic) is a Russian broadcaster owned by Gazprom. It is the only private Russian broadcaster. NTV operates the NTV Plus satellite broadcasting service which uses several satellites including the Eutelsat W4 satellite at 36.0°E and Yamal-200 at 90.0°E. Combined with hundreds of terrestrial stations, this allows for coverage of most of Russia. NTV also broadcasts via its **NTV Mir** affiliate using a combination of satellite feeds and cable TV to an international audience.

15. Installing the Dish or antenna

This chapter considers the practical issues involved in installing a terrestrial or satellite antenna and connecting it to a receiver. In addition to the discussion here, it is also important to read the instructions provided with any new antenna. Those instructions may include specific details about the particular antenna and/or information that preclude the general suggestions here. On the other hand, some antennas only come with very general installation instructions that omit some of the guidelines given below.

Installing a terrestrial antenna

In Chapter 8 we discussed the different types of electromagnetic frequencies used for broadcasting. Before buying or installing a terrestrial antenna you need to figure out which bands are used for broadcasts in your area: UHF or VHF. In most urban areas both are in use. Based on this information, the distance to the broadcasting towers (and hence the signal strength), the directions of the different broadcasters, and the information in the chapter on antennas, you can select one or more antennas. The rest of this chapter is relevant to both terrestrial and satellite antennas, but it places more emphasis on satellite dishes since their placement issue is usually more delicate and complicated.

Coaxial cable is the preferred way to connect an antenna to a receiver, especially in the case of an outdoor antenna. Twin-lead ribbon cable is also possible for terrestrial antennas, but it is more susceptible to interference and signal loss.

Selecting the location

The paramount issue to consider in installing a dish or terrestrial antenna is the placement of the antenna on the property. In the case of a dish antenna, we will assume that the goal is to pull in reception from a single geostationary satellite, although the issues for installing an antenna with a motor drive that can be moved between satellites are about the same. The key issue is aiming the antenna at the source of the signal using an unobstructed line of sight. When installing an antenna to receive terrestrial broadcasts the issues are roughly the same as for satellite broadcasts, but the need for a direct unobstructed line of sight is less critical.

As we saw earlier, all of the geostationary satellites in earth orbit lie over the earth's equator. This means that for a person in the northern hemisphere (including the USA, Canada and Europe) the satellite dish needs to have a view of the southern sky. For people in the southern hemisphere, the situation is inverted and a view of the northern sky is required. If you happen to live right on the equator, you just need to look upwards along a line going east and west. Naturally, the dish will have to be aimed at a specific point in the sky, given by its direction (also known as azimuth) and elevation. The precise direction and elevation needed to aim at a specific satellite depends on exactly where the satellite is along the equator (i.e. which orbital slot it occupies), as well as the latitude and longitude of the dish antenna on the earth's surface. A table of these values is often supplied with a dish antenna, or else is accessible as a built-in function on some receivers. In addition, many web sites will do the satellite lookup and direction calculation for you. Such web sites can usually be found using a web search engine using a search string such as "**satellite pointing calculator**" or via the web site associated with this book (http://www.Y1D.com/DTVbook/links). Only a rough estimate of your latitude and longitude is required to determine this pointing direction, and you can find this using a GPS receiver or else turn to the web as well (for example pick a large city nearby, let's say Chicago, and search for "Chicago latitude longitude").

A good location for a dish antenna ought to have a clear view in the required direction. Keep in mind that foliage may change during the year, so a view that is clear in winter could become blocked by leaves in the summertime. If you are putting in place a very permanent fixture, you may also want to consider the places where tress may eventually grow. Some locations may be more susceptible to wind than others, and wind can cause vibration (which leads to reduced reception). Make sure that the entire dish will fit in the desired location (i.e. it is not too close to a chimney), that the coaxial cable from the dish can be safely led back to the receiver, and that a ground can be connected (a ground wire is important). Stand at the selected spot and look in the direction the dish will be pointing to assure a clear line of sight: the Internet abounds with stories of people who installed a dish antenna and then discovered the line of sight was obscured.

For a terrestrial antenna mounted on the roof, the antenna should be four feet from the apex of a peaked roof, since even a wood roof can affect the signal. For homes with a metal roof, a good rule of thumb is that the actual antenna element should be ten feet or more above the roof, achieved by using a sufficiently long mounting pole.

Installation procedure

A critical issue is mounting the antenna on a firm, rigid structure that will not vibrate in the wind, nor fall over. The most common location for a terrestrial outdoor antenna or a home satellite dish, especially for C or Ku band, is on the roof. This often provides the best line of sight. On the other hand, roof mounting can make installation more difficult. The issues one needs to be wary of include the risk of causing roof leaks and that fact that a poor placement can make maintenance (such as debris or snow removal) more difficult. Mounting the dish on the side of a building has similar issues, but it is often the only option for apartment dwellers. For apartment dwellers a balcony is often an option as well, although this can easily lead to vibration problems. For those with suitable yard space, mounting a dish antenna on a metal pole that is anchored in a cement pad may be the best solution with respect to technical issues.

Note that in many regions landlords are legally obliged to allow an apartment dweller to install a satellite dish in any area where they have exclusive access. In the United States this is based on federal legislation from the FCC called the **Over-the-Air Reception Devices Rule** ("**OTARD**"). Similar laws exist in parts of Europe. This right may even override prohibitions that are explicitly set out in a lease, depending on the local laws.

The next step after selecting a location (and perhaps putting a mounting pole in place) is to assemble the dish hardware. This is usually a routine mechanical job that is simple in the case of a DVB antenna, but which can be complex for large C-band dish. Likewise, terrestrial antennas can vary in the complexity of their assembly.

When connecting the cables to the antenna it is important to ground the antenna. This can usually be done by leading a cable from the antenna itself to a grounding point, which can be a household grounding cable or even a deep stake in the ground. If you are uncertain about what constitutes a good ground, you should seek an expert opinion. Grounding serves two important purposes: it can provide an alternative electrical path if the antenna is hit by lightning, and it allows static electricity which can accumulate naturally to drain off. While lightning is the most spectacular risk you might want to avoid, damage caused by simple static buildup is probably a more common problem.

Positioners

Some viewers want to be able to pull signals in from a variety of television broadcasters, or a set of satellites, without having multiple antennas. This is especially true for those using C-band satellite dishes, and for those receiving terrestrial broadcasts from several different stations.

For terrestrial antennas, a small rotor can be mounted on the mast that holds the antenna(s). A cable leading to a control unit inside the home connects to this rotor and it allows the antenna to be rotated to aim in different directions.

The same principle is used for satellite dishes, but the apparatus is a bit more complicated. A positioner arm (or actuator) is attached to a motor and used to rotate the dish horizontally so that it can point at different satellites. Various alternative arm designs are available which provide different amounts of turning range. These arms are most popular for large dishes. For smaller dishes used with DVB it is usually cheaper and easier to put additional LNBs on the dish, and/or to mount additional dishes (see Figure 34). Some elliptical dishes are specially made to span a large range horizontally and allow many LNBs to be arranged in front of them, thus making a rotor less desirable (Figure 33).

Multiple receivers

Some people want to connect multiple receivers to a satellite feed (or to a group of satellites). This can be convenient if there are televisions in different rooms that each need access to the signal from a satellite dish. Even if only one satellite dish is being used, it is generally not possible to simply connect the cable from the dish to a mere splitter and feed both receivers. This is because most LNBs need be sent control messages to select the polarization of the signal associated with a specific channel. Using a simple splitter will usually mean only half the channels are visible to one of the receivers, and/or that channels go off on one receiver when the other one changes channels.

The standard solution to this problem us a device called a **multiswitch**. This is a special form of switch that is specifically built to connect to an LNB that has two independent outputs (or to a pair of LNB's). The multiswitch permanently sets one of the connections to the LNB to a single fixed polarization (e.g. horizontal) and the other connection to the opposite polarization. This way it can access either polarization. Then it uses a switch to provide either one of these signals, without changing any settings on the LNB, and a splitter to provide any of the two signals to as many receivers as it supports. A typical multiswitch supports two LNB connections and 4 receivers.

Multiple dishes or LNBs: DiSEqC

It is now common for a single broadcaster to supply data using more than one satellite for their customers. This typically implies the use of multiple LNBs in front of a single dish, but it may even require the use of multiple dishes. This is even more likely when people want to listen to free-to-air programming which is present on a diverse set of satellites. Typically one uses a single cable from the receiver to the rooftop and a switch there to select the correct output to send down. The receiver then sends a message up the cable to this switch, which selects which data source (e.g. which polarization) to pass down. Many **compound LNBs** include multiple internal antennas for different polarizations and have a switch built in.

The standardized protocol used for this kind of switch is called **DiSEqC** and the switch is typically called a **DiSEqC switch**. DiSEqC stands for Digital Satellite Equipment Control and is usually pronounced "die-seck". The protocol is administered and was developed by Eutelsat, who have trademarked the name, but the protocol is openly documented and free for use by anybody.

Figure 46: DiSEqC switch, 4 satellite inputs to one receiver.

There are various reasons for sending commands from a satellite receiver back to the LNB, the dish, or some intermediate switch. The DiSEqC protocol can also be used for cable TV equipment. It is a protocol that provides a low-bandwidth bi-directional digital data bus, piggybacked on the same cables used for video data transmission. The DiSEqC protocol is based on having a **master** device and a set of **slave** devices that receive its commands, using a network connected as a tree (i.e. in a tree **topology**, see Fig. 47). The commands that control the settings of the slaves are sent by the receiver, which plays the role of the master. Some receivers allow the user to directly manipulate the DiSEqC programming, while others keep it hidden within the built-in firmware.

The DiSEqC network operates by having all commands originate with the master device, and the slave devices either respond with information, or change their behavior in response. Slave devices must also pass incoming DiSEqC messages along the tree so that downstream devices also get the commands from the master, and the master gets downstream information sent back from lower-level slave devices. Every slave has an ID number, an address associated with it, and for the simplest and most reliable operation these addresses should be unique. In practice, the "network" is often made up of just a receiver as the master with the slave being a single switch connected to a pair of LNBs.

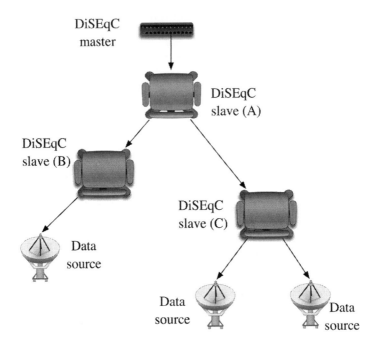

Figure 47: Example of a DiSEqC network.

The messages are sent along the cable using signals at 22kHz, a frequency corresponding to an audible tone, and with a maximum data rate of 666 bits per second. This frequency was used because 22kHz signaling was also used for an older, simpler **tone burst** system for controlling dishes. Another legacy system for signaling is based on switching between two voltages, at 13 and 18 volts. This voltage-based system is also supported by DiSEqC switches. Both of these legacy schemes (tone and voltage) are implemented by sending strings of either nine binary '1's or nine binary '0's using the DiSEqC coding. In the case of binary ones, this sustained pattern will correspond to a 22kHz tone and a change in the average voltage, thus having the same effects the old-style control signals. A DiSEqC slave will respond to old-style signaling until it receives a genuine DiSEqC command, and which point a slave is supposed to change to DiSEqC mode. The voltage for DiSEqC tones is roughly 650 mV (peak-to-peak), and the maximum current a slave device can draw is 500 mA.

DiSEqC slave devices are always connected in a network tree that eventually leads back to the master, the receiver. The master is connected to the first slave in the network. Additional DiSEqC slaves can then be connected to the outputs of this first slave. The DiSEqC term for a slave connected to another slave like this is a **cascaded switch**. In order for slaves further down the tree to get the messages, the slave above needs to pass on any messages it gets. Some of the messages, however, can actually turn the downstream slave's electrical power on and off. Further, when a slave gets turned on it usually needs a short time to start running properly (measured in fractions of a second). As a result, a message is often sent more than once if it needs to propagate to a slave that might be further down the tree, and thus which might have just gotten turned on due to the message. A general rule of thumb is to repeat a message a number of times equal to the depth of the tree (e.g. twice for the network shown here). That way, if a slave gets powered-up as a result of a message, the message will be repeated so it has a chance to propagate through this slave after is has been turned on and "warmed up".

The time needed to send a DiSEqC command is typically no more than 5 milliseconds. When a message is repeated an additional 50 milliseconds of delay is needed between the repeats, and thus, each repetition takes about a tenth of a second. As a consequence, using too many repeats can lead sluggish channel changing.

The DiSEqC protocol is composed of short messages. These messages include commands that are each one or two bytes long. The language constituted by these messages allows devices to have 8-bit ID codes, although in general actual switches can only take on a very limited set of different IDs. The commands describe how a small repertoire of control bits in the slaves can be manipulated. These control bits have functional names like "frequency" and "hi band/low band" which presumably is meant to make them easier to understand.

There have been several versions of the DiSEqC protocol and it also has three **DiSEqC levels**: 1, 2 and 3. DiSEqC also has different versions. For many consumers using DiSEqC to control DVB LNBs, however, the distinction between the versions is not significant. Level 1 of the protocol has version numbers of the form **1.x** (e.g. 1.0, 1.1 and 1.2) and uses one-way communication from the master to the devices attached to the slaves, but to not back. DiSEqC level 2 (versions 2.x) allows for two-way communication between the receiver and the devices at the end of the network tree (i.e. the leaves). DiSEqC version 3.0 is much less common, but supports "external bus control", meaning the master (i.e. receiver) itself can be programmed over the bus.   Versions of the protocol where the last digit is at least 2 (i.e. version 1.2 an 2.2) allow a single-axis positioner to be controlled as well, for example to change the horizontal aiming direction of a satellite dish using a positioning rotor.

DiSEqC messages have fairly rigid structure. Each begins with a framing byte that can indicate if a response is required or not (for example **0xE0** is a start-of-message byte with no reply expected). This is followed by an address selector that specifies which switches in the network should respond to the message (for example **0x11**).   Next comes a command byte and then, depending on the command, a data byte.   Standard device families have specific standardized addresses. LNBs have the address **0x11**, switches with DC current blocking have the address **0x14**, switches without blocking have address **0x15**, SMATV switches use **0x18**, etc.   In addition, all devices should respond to the **wildcard** address **0x10** (to be used in simple networks with just one switch). Furthermore, there is a command that tells a device to change its address, which is useful in multi-level networks to change the addresses of devices that might get confused with one another.

Some examples of DiSEqC commands follow.   Here we assume an LNB having an internal DiSEqC switch with address 0x11 is connected to an external switch with address 0x14.

Common DiSEqC messages in hexadecimal are shown below:

"Power On (any device)" : **E2 10 03**

"Select switch position A" (address 14h): **E2 14 22**

"Select switch position B" (address 14h): **E2 14 26**

"Select switch position x (0-15), Committed switches": **E0 10 38 F**x

"Select switch position x (0-15), Uncommitted switches": **E0 10 39 F**x

"Select switch position 6 on switch 0x18": **E2 18 39 F5**

Several internal flags in the DiSEqC specification have specific meanings attached to them, at least when they are **committed** to their official functions. DiSEqC switches can also be **uncommitted**, meaning they treat these bits as arbitrary values that are joined together to form a 4-bit number. In the uncommitted case, the bits are labeled **SW1** through **SW4**. These bits can be individually manipulated, or set all at once. For example, the command 0x29 sets SW1 to a binary true (1) value, while the command 0x39 0xC8 sets all four switches at once to the values: SW1=false, SW2=false, SW3=false, SW4=true.

The **committed** bit flags (and the internal hardware Control Pins in a switch) are labeled as follows:
- Local oscillatior (L.O.) frequency (low=0, high=1)
- Polarization (vertical=0, horizontal=1)
- Satellite position (A=0, B=1)
- Option Switch (A=0, B=1)

Since the special address **0x10** can be used to control any SMATV switch, generic switch, or LNB, a few simple commands are especially common based on this value. By using this address, we can construct a set of messages to change between the inputs on almost any 4-input switch in a simple setup (based on controlling SW3 and SW4):

Basic DiSEqC messages for network containing only a 4-input switch

Switch position	Message string (4 hexadecimal bytes)
1	E0 10 38 C0
2	E0 10 38 C4
3	E0 10 38 C8
4	E0 10 38 CC

As we saw, prior to the widespread adoption of the DiSEqC standard, several types of alternative switch technology were developed using the legacy voltage or tone-based signaling mentioned above. In general, these are much less flexible than DiSEqC, but work well for simple networks. Since DiSEqC devices can inter-operate with tone-based equipment, the two can be used together so long as the legacy devices are placed after the DiSEqC switches in the network, since the legacy switches will probably not pass on the DiSEqC messages that they do not understand.

There are also some non-standard variations. For example, Dish Network once produced a variety of switches called the **SW-21**, **SW-44** and **SW-64** that used tone-based control to switch between either 2 , 4 or 6 inputs (LNBs) respectively. The last digit of the model number on these common switches refers to the number of receivers that can be connected to them. These switches are widely used, but they are reputed not to inter-operate well with other switch technologies, presumably because they have proprietary features and non-standard characteristics. (More recent Dish Pro switches such as the **DP34** and a Dish Pro Plus **DPP44** are based on the DiSEqC protocol.)

Some LNBs, such those used for Dish Network's "Dish Pro" models, also include circuitry to simultaneously receive horizontally and vertically polarized signals, and return both types polarizations at the same time without a switch. This **bandstacking** is accomplished by mapping signals with horizontal polarization into new unused frequencies, so that they appear as if they were additional transponders that have vertical polarization. Thus, from the point of view of the receiver, the LNB with two sets of polarizations appears to be a single LNB connected to a satellite with twice as many transponders as usual, all using vertical polarization.

If such a bandstacked LNB is used for generic equipment (e.g. for FTA programming), then the receiver needs to be configured to look for channels that have horizontal polarization on non-standard shifted frequencies.

Using a signal meter

A signal meter is a small device that you can insert between the dish antenna and the satellite receiver. It usually has one coaxial cable connector that leads to the dish, another that leads to the receiver, a sensitivity adjustment knob, and a large meter that indicates the signal strength from the dish antenna (see the Fig. 48). It is an utterly indispensable tool for anyone who wants to align a satellite dish, especially given its very moderate price (typically about $25).

The signal meter passes the antenna signal from the dish back to the receiver, but it also displays the strength of that incoming signal on its front panel. In addition, the majority of signal strength meters also produce an audible tone whose loudness or pitch changes in proportion to the measured signal strength. The adjustment knob allows the meter to be tuned to be more or less sensitive to a given signal.

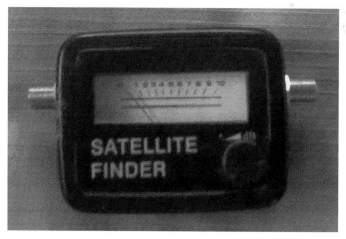

Figure 48: Signal strength meter

The meter can usually be used right beside the dish antenna (usually with a very short cable between the dish antenna and the meter), for example up on the roof of a building. It allows a person to adjust the direction of the dish antenna and immediately see the effect on the received signal strength. The response of the meter is far faster than the signal strength indicator displayed by a receiver on the TV screen. It also avoids the terrible hassle to attempting to see a television set located inside the house, while adjusting something on the roof, or shouting to an assistant. One warning, however, is that unlike a receiver it cannot distinguish between the signals from different satellites.

Usage Procedure

The procedure for using a satellite signal meter is as follows. First, using the knob adjust it so that the meter just barely starts to rise in whatever direction the dish in currently pointing (initially in some arbitrary direction, for example). Then adjust the aiming direction of the dish attempting to make it point more accurately at the satellite of interest, and watch the meter while doing this. Move the dish antenna so that the displayed signal strength increases as much as possible. As the dish points more accurately at the satellite it will be necessary to use the adjustment knob to reduce the sensitivity, to avoid having it go right off the scale. It may be necessary to reduce the sensitivity of the meter repeatedly as the aiming improves more and more. The whole procedure works very well even if the initial guess for the satellite position is quite poor.

16. Wiring and Connectivity

All the components of a digital television system need to be wired together. While this sounds trivial, it is complicated by the fact that different parts of the system need either specialized wiring, specialized connectors or have specialized signal-carrying characteristics.

Antenna connections

Satellite and terrestrial antennas each need specific types of cable. We will consider each in turn. A common failure for novices installing their own equipment is to use the wrong cable type to connect to the antenna.

Satellite television

The same connection used to carry the signal from the LNB at the dish down to the receiver is also used to carry switching messages back from the receiver to a switch (sometimes within the LNB). The video signal coming down is at a high frequency while the control signal is at a relatively low frequency. In either case, this suggests that the cable needs to have good performance at high frequencies and acceptable performance at low frequencies as well.

The connection from the satellite dish to the receiver should be made using **RG6** coaxial cable, a standard grade with known performance. This is cable with a center conductor made of either solid or stranded wire, an insulating material (dielectric core), an outer electrical shield, and a jacket. The wire is usually copper or copper-clad steel. The dielectric insulator between the inner wire and the outer shield is what determines the high-frequency properties of the cable. Any insulator is characterized by its dielectric constant, which describes to how it reacts to voltages across it. The outer shield is a combination of wire braid and metallic foil, such as aluminum. Video quality cable needs to have 59% coverage by the braid, as well as being 100% covered by the foil. Less coverage means more signal loss.

Video-grade cables have the prefix **RG** for **radio grade** and vary according to impedance and frequency response. Cables generally are electrically characterized by two values: their signal loss and their impedance, where impedance is often the more important of the two. Impedance is a function of the spacing and material between the two conductors making up the cable and does not depend on the cable length. Signal loss depends on the length of the cable. RG6 cable in particular has an impedance of 75 Ohms. Cable with the wrong impedance leads to electrical reflection at the ends; the signal essentially bounces off the end, and thus there is signal loss. Although the cables that are used for cable television (**RG59**) or old **10-base-2** Ethernet look about the same as RG6 cable from the outside, they have serious losses at the higher frequencies used with satellite broadcasts. As a result, old RG59 cables should not be used, except maybe for very short runs where a lot of signal loss per unit distance won't be noticed. Coaxial Ethernet cables, like radio antenna cables, also have the wrong (50 ohm) impedance for video applications and are usually RG8 or RG58. Even at low frequencies, RG59 cable has higher resistance thab RG6. At DC an RG59 cable has a resistance of about 16 ohms per 1000 feet (304 m) while RG6 has a DC resistance of about 8 ohms per 1,000 feet.

The coaxial antenna cable is terminated at each end by an **F-connector** that is composed of a round screw-on outer enclosure with the center wire protruding from the middle (see Figure 49). The connection between the F-connector and the cable needs to be waterproof on the outdoor end. These connectors can be applied to raw cable quite easily with no special tools by using a commonly available twist-on connector. An inexpensive crimping tool will provide a more robust connection, it will also be cost effective if more than just a couple of ends need to put on.

Figure 49: An F-connector.

RG11 cable has slightly better characteristics than even RG6 cable. The DC resistance is lower and the cable is more physically robust, but it is a bit thicker than RG6 and thus harder to manipulate. It's probably not going to matter except for very long runs of cable.

Analog video interconnects

Transmitting the video data from a conventional television receiver to other components of the system is the easiest and most conventional cabling problem. In general, standard video ("RCA") cables can be used, and these are relatively inexpensive and readily available. They don't provide the best possible quality, but they are the common denominator for standard definition video and audio.

Component, composite, YUV

Component video refers to the separation of a video signal into three channels, each carried on a separate physical cable, plus some additional cables for synchronization data. As seen earlier, one variation of this is YCrCb signaling, while another is RGB (these are two ways of separating the color signal into three different parts).

High-quality analog systems sometimes include RGB interconnections. With an RGB signal, a synchronization signal (a clock) is required as well. This sync signal is often sent on the channel used for the green color component (called **sync on green**), or it is sometimes also present on all three-color channels. It is also possible for the sync signal to be transmitted on a fourth extra wire, or horizontal and vertical sync each to be transmitted on individual wires (which provides the highest-quality RGB analog interconnection). These different schemes lead to component RGB connectors having either 3 wires (sync on green), four wires, or five wires (independent horizontal and vertical sync). These connectors can be equipped with RCA phono connectors, or BNC connectors (see Fig. 50). BNC connectors are more common for devices catering to the professional or commercial video market.

Figure 50: BNC cable connector sometimes use for component video.

Component video, being intrinsically analog, will gradually degrade as the length of the cable increases, and will usually result in a higher quality picture if the cable quality is good. Runs of up to 50 meters (150 feet) are possible with no noticeable signal degradation, but require great care as this length limit is approached. In practice, component video cables are often kept under 2 meters (6 feet) in length. Moderate lengths of component video cable can provide display quality that is indistinguishable from that provided digital video cabling, but many modern devices intentionally cripple the component analog outputs to try and plug the analog hole. As a condition of licensing for several key HDTV technologies, devices are required not to provide analog output at resolutions substantially above standard definition (either by shutting analog output off for HDTV content, or by providing it only at reduced resolution).

Whereas composite video uses only two wires, **S-video** (Y/C video) uses an additional wire to carry the video signal, and thus keep the Y and C (brightness and color) components of the video signal better isolated from one another. This leads to a potentially better signal than traditional composite video. The S-video cable and connector come in two variants: **S1** and **S2**. The S1 connector has fewer wires (just 4) while the 7-pin S2 connector adds an extra signal to distinguish the aspect ratio of the image (4:3 for standard definition versus 16:9 widescreen). The **S3** variation also includes a signal to indicate if the video in a letterbox format.

Firewire

Firewire is a standard for both wiring and data exchange developed by Apple Computer and standardized officially as IEEE 1394. It is a general-purpose exchange standard for digital data, and video data is just one specific application for it. It is supported directly by many computers and digital video cameras, and assorted additional devices. The most common cables for Firewire use 6 pins and can transmit both data and electrical power (although only for devices with limited power demands). 4 pin cables also exist, and can interoperate with 6-pin devices (so that 4 to 6 pin cables are also common), but 4-pin cables do not carry power. A high-speed variation of Firewire called **Firewire 800** also exists and uses different cabling, but essentially the same data transfer protocol.

A specific copy protection and rights management layer has been developed for video data transferred over Firewire (see the discussion of DTCP on page 184).

Serial digital interface (SDI)

Serial digital interface (**SDI**) cables are defined as part of the **SMPTE 259M** standard for video transmission (also defined as the international standard **ITU-R BT.656**). It deals with the transmission for digital standard definition video. Cables used for SDI are coaxial, use BNC connectors, and are typically used only for professional broadcast-quality video equipment. SDI video is typically uncompressed and hence has a very high data rate. There is a closely related standard for high-definition video (**HD-SDI**), but due to the even higher data bandwidth, the cables for HD-SDI are often fiber-optic.

Because SDI video is unencrypted, typically not copy protected and yet of very high quality, it is generally only used in equipment that is not destined for the consumer market.

High-Def Media Interface (HDMI) and DVI

HDMI and **DVI** cables are two different multi-conductor high performance cable types used for digital video. Both **High-Definition Media Interface** (HDMI) cables and **DVI** cables transmit digitally encoded video signals for display on a monitor, although HDMI cables also carry audio signals. The signals are often encrypted using HDCP. An HDMI cable has 19 conductors within in. When DVI cables are used, separate cables are needed for audio. Both HDMI and DVI cables deliver the video using **Transmission Minimized Differential Signaling** (TMDS) as an electrical protocol, which explicitly encodes image information using RGB encoding. Since both types of cable carry the same electrical signal for video, cables that convert from one connector type to the other are straightforward.

While digital signals should, in principle, cope well with noisy transmission, standard HDMI/DVI cables are not very robust with respect to signal loss. This is partly due to the high frequencies being transmitted as well as the tight packaging of the conductors within the cable. As the cable length increases the bits being sent will gradually be harder for the system to resolve, and at some critical distance the signal will become unrecoverable.

There is usually a short intermediate range where only a few single pixels are being lost, and odd artifacts and sparkles will appear in the image, but this intermediate range only covers a small range of distances. As distance increases, digital signals will experience fairly sudden and complete failure, as illustrated in the figure below. Your video may be work well with an 8-meter (24 foot) cable, and then give no picture at all with a cable of 12 meters (36 feet). In comparison, analog signals gradually get worse and worse with cable length, but don't vanish completely for much longer, and the errors that do show up with digital signals are often more distracting. Universally reliable cable lengths for HDMI and DVI are 5 meters (15 feet) or less. Even so, HDMI cables should be treated with care and should not be bent or crimped severely due to the very high frequency signals they need to conduct and the close spacing of the internal wires.

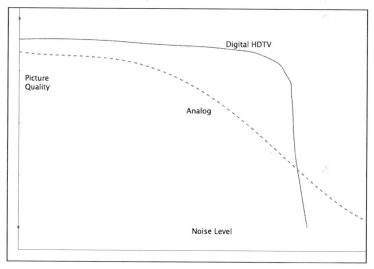

Figure 51: Picture quality degradation as a function of distance. Digital is much better when it works, but fails more completely when things get too noisy.

In addition, HDMI cables come in two different performance grades: category 1 and category 2. Category 2 cables have higher performance (in terms of bandwidth) and are needed for full quality transmission of signals at 1080p resolution. Category 1 cables, which are by far the most common, are suitable for 1080i video or lower resolutions.

The data specification for HDMI cables also comes in more than one variation, notably version 1.2 and 1.3. Version 1.3 of the protocol has not been used extensively as of 2008, but has higher maximum bandwidth and is intended for 1080p signals and beyond. It is also designed to support more bits of color information and very high performance audio signals. The color data includes support for 10 to 16 bits of color information per color channel, as opposed to the 8-bit color used with HDMI version 1.2.

SCART cables

SCART connectors have 21 pins and carry video signals as well as audio. SCART is a European standard for analog video only, not for digital video. It was very widely adopted and standardized, and most European analog video devices support a SCART connector. Oddly enough, the cable was designed in France and the name was originally an acronym for Syndicat des Constructeurs d'Appareils Radiorécepteurs et Téléviseurs (although that is merely a historical curiosity), but in France the cable is typically called a **Peritel** cable. It is also sometimes referred using the less common terms **Euroconnector**, or **EuroSCART** connector. The official standard is **IEC 60933-1**.

It supports NTSC, PAL and SECAM signals. SCART cables always carry composite video (as a kind of lowest common denominator), but can also carry additional video signals as well as audio. The cables can carry three different variations of the signal, differentiated by what kind of video signal is carried in addition to composite. SCART 1 carries composite video as well as RGB video. SCART 2 carries composite video and S-video. SCART 3 carries composite video only. All these variations look the same from the outside, which can sometimes lead to confusion regarding what is actually available.

Most SCART-equipped video devices, including television sets, provide video output on the SCART cable. This feature is sometimes used in decoding/descrambling devices that receive a scrambled signal on the SCART cable from a TV tuner, and then return a modified or descrambled version back on the same cable, which the TV then displays. A side effect of this combined input and output on the same cable is that SCART cables can sometimes lead to ghost images due to interference between the different signals. Pin 19 on a SCART cable is used for composite video output and the physical removal of this pin can sometimes alleviate this kind of ghosting problem (assuming that no video output is required on the cable).

SCART connectors are typically bi-directional and allow both video and stereo audio to be transmitted as well as received. In general the cables are heavy and stiff (due to the fact they carry 21 individual wires, and in good quality cables each video line is a miniature coaxial cable), so that although the maximum practical length is up to 15 meters (45 feet) such long cables can be expensive and cumbersome.

Figure 52: SCART connector.

Figure 53: SCART input jack on the back of device.

Audio connections

One of the major audio interconnection methods is called AES3. Because it is defined by the **AES/EBU** standards bodies, it is often referred to by the name of the organizations themselves, where the **AES** and **EBU** acronyms stand for the relevant American (US) and European standards organizations (the Audio Engineering Society and the European Broadcasting Union). This connection system was developed to transmit linear **pulse code modulated** (linear PCM) audio, which is a form of uncompressed digital audio similar to that found on ordinary CD audio recordings. **Linear** refers to the fact that numerical values are encoded in proportion to their strength (i.e. linearly, as opposed to logarithmically) and **pulse code modulation** refers to the encoding as a sequential series of audio energy pulses. It is based on sending stereo data in a sequence of packets called **frames**, and the encoding accuracy is between 16 and 24 bits per audio sample, at a rate of between 22.05kHz and 192kHz. The AES3 standard is heavily focused on physical interconnection standards and can be used with **XLR** (balanced) connectors, coaxial cables with RCA connectors, or optical cables with F05 connectors.

S/PDIF (Sony/Philips Digital Interconnect Format) is a digital audio transmission protocol developed by the Sony and Philips Corporations, which has been absorbed into the AES3 specification as a variant. It is used to carry linear PCM AES3 signals, but is also able to transmit compressed audio and 5.1 channel surround sound, as well as just stereo. It seems to have been developed specifically for consumer use, but is also used in some professional equipment. S/PDIF is also associated with the international standard **IEC 958 type II**.

Another extension of AES3 called AES3_SMPTE (standard SMPTE 340M) describes the transmission of other data, and specifically non-PCM audio that is compressed using the AC3 standard (discussed on page 30).

Appendix A.   Buyer's phrase book

In this section we provide a simple translation guide for the feature lists seen on home theatre and digital television gear. This translates the technical jargon into a more comprehensible form, and should help make selecting components a bit easier. Of course, a fuller description of the jargon is provided in the rest of the book.

Television sets

Feature name	English Translation	Importance
Anti burn-in	Prevents a single image from staying on-screen too long.	Many modern displays will not burn in, and this kind of static image is usually only produced by gaming consoles.
Anti-glare	Coating on the screen to reduce reflections.	Can be useful in brightly lit rooms, but it may slightly reduce contrast.
Freeze frame	Allows an image on the screen to be frozen like a photograph.	Of debatable value.
Electronic program guide (EPG)	Captures and displays a text listing available programs for one or more days.	Very convenient..
Integrated DVR	Digital video recorder included.	Very significant.
Wi-Fi	Wireless connection to computer or other devices.	Can be very useful, depending on what it is used for.

Feature name	English Translation	Importance
Hard drive	Included disk for storage. Usually only useful for DVR functions, in which case as much space as possible is desirable (20Gig is small, 300Gig counts as fairly large).	
Ambilight, Ambisound, AMBX	Proprietary Phillips technology to enhance picture quality under changing brightness on sides of image.	Of debatable value.
Plasma display	Screen display technology.	The tradeoff between plasma and LCD is complex and subjective.
LCD display	Screen display technology.	The tradeoff between plasma and LCD is complex and subjective.
Flat screen	The screen is flat.	Standard feature present on all modern televisions.
Channel Labeling	On-screen text providing channel ID.	
Closed Captioning	Printed transcript at bottom of screen that allows spoken content to be read.	Useful for hearing impaired, or those who may watch with volume set ery low.

Feature name	English Translation	Importance
Parental Control	Automatic censoring of shows with certain ratings.	Very useful if young children will watch without close supervision.
Cable Card Slot	Allows CableCARD to be inserted to avoid the need for an external cable TV box.	Only supported in the USA. Some cable boxes may provide more functionality than a cableCARD.
Coaxial Cable Inputs	High quality analog input.	Only useful for those with high-end semi-pro analog equipment.
A/V (Composite) Inputs	Standard analog input.	Important, but universally available.
Component Video Inputs	High quality analog input.	Only useful for those with high-end semi-pro analog equipment.

Satellite and Cable TV Receivers

These are features typically listed on satellite or cable receivers, particularly Free-to-Air receivers. Receivers that are tied to a single provider, such as DirecTV or Dish Network, typically have a more standardized feature list with fewer possible options.

Feature name	English Translation	Importance
EPG (Electronic Program Guide)	Provides a guide for what's on now and in the future.	Standard feature present on all modern receivers.
Support for MPEG-2 Digital & DVB broadcasting	Works will normal types of digitally encoded broadcast.	Standard feature present on all digital satellite receivers and digital cable (MPEG2).
950 - 2,150 MHz input Frequency (IF Loop Throughout)	Kind of frequency used to communicate with dish antenna LND.	Standard feature present on all modern receivers.
Variable Input Symbol Rate (1-45 MBps)	Class of signals that can be decoded.	Standard feature present on all modern receivers.
Quick Channel Change	Changes channels quickly.	Not very notable, vague.
5,000 Channels TV & Radio Program	Supports a very large number of channels.	Commonplace, not notable, but a big number is good.
Universal Remote	Remote control can be set to control TV or other devices as well.	Commonplace, such remotes are fairly cheap, but not ever receiver has one. Usually universal remotes included with a package are not as good as what can be bought off the shelf.

Feature name	English Translation	Importance
Flexible reception of SCPC & MCPC from C/Ku-Band Satellites	Various satellite signal encodings. Support for big C-band dishes is stated explicitly.	Standard feature present on all modern receivers. Few people want C-band today.
S/PDIF (Digital Audio or Dolby AC3 Stream out)	Digital audio output.	Digital audio out can be desirable if you want to attach a high quality sound system.
Timer Function (Automatic Turn On/Off by Setting Function)	Can turn on or off automatically.	Can be used in conjunction with a recorder to make automated recordings. Pretty common, but important.
Multi Lingual On-Screen Display or Multi-Language GUI	On-screen menus can be shown in choice of languages.	Standard feature present in some form in all modern receivers. Specific languages may be important to some viewers, though.
PAL/NTSC Conversion	Can receiver analog video in both North American and European formats.	Notable.
256 Colors (Graphic User Interface) or 8-bit color interface	The menu display is shown with many colors.	Standard feature present on all modern receivers.
PIG (Picture in Graphic)		
N Favorite List	Stores a list of favorites channels for easy access.	Standard feature present on most modern receivers.

Feature name	English Translation	Importance
Channel Editing Functions	Allows the on-screen list of channels to be sorted. Allows channels to be deleted from the menu of what is shown.	Standard feature present on most modern receivers.
Parental Lock, Parental Control	Allows channels or programs to be locked out or hidden using a secret code.	Commonplace. Can be useful to prevent children from accessing inappropriate content.
Move, Delete, Favorite Edit	Allows the on screen list of channels to be manipulated.	Standard feature present on most modern receivers.
Manual / Satellite / Network Scan PID Scan Transponder scan	Allows list of available channels to be determined automatically by inspecting the satellite signal itself.	Important feature for those wanting to access FTA or wild feeds.
Satellite Scan - Simultaneous Scan with 4 Satellites	Allows faster scanning of satellite signals.	Timesaving feature, but only likely to be used infrequently (when system is installed).
DiSEqC 1.3	Multi-satellite switching signals.	Standard feature present on all modern satellite receivers.
Firmware Upgrade via RS-232	Uses a computer connection to upgrade the software. RS232 is a particular cable type that may not be present on all computers.	Software upgrade is useful. RS232-to-USB adapters are inexpensive.

Feature name	English Translation	Importance
Firmware Upgrade (via SD-Card or other medium).	Can upgrade internal stored operating program.	Software upgrade is useful. Doing this via a memory card, such as used by digital cameras, is convenient assuming your computer supports it.
Firmware Upgrade via Ethernet or IP or Internet or TCP	Uses an internet connection for software upgrade delivered from manufacturer.	Software upgrade is useful. Internet-based upgrade is usually the most convenient.
Component Out	High quality analog output.	Useful if you want an analog feed, and your television supports component input.
Composite Out	The most basic analog TV connection.	Standard feature present on most modern receivers.
HDTV support	Output in digital format.	Important if your television accepts it.
HDTV out, DVI Out, or HDMI out	Output in digital format (same as above).	Important if your television accepts it.
Conditional Access Module or CAM support or Access card	Supports proprietary encryption schemes.	Important for those who seek to use proprietary encryption cards (e.g. some European viewers). For North Americans, proprietary encryption schemes are only legally supported by the manufacturer's own equipment.
RTC/ Real Time Clock	Has a build in clock.	Standard feature present on all modern receivers.

Feature name	English Translation	Importance
Secure Digital (SD) Card Access	Includes a slot for a SD memory card to allow display of digital photos, or upgrades	Generally not a key feature for most users.
MP3 Audio file decoding	Can play digitized audio files.	Can be useful if you regularly create digital CDs
Upgradeable software or Firmware upgradeable.		Useful feature. Common, but not universal.

Appendix B. Satellite problems

Common sources of poor satellite reception

Weak satellite reception can lead to a complete loss of picture, or in borderline cases it can lead to poor picture quality and a blocky picture with rectangular rapidly moving artifacts. This blocky degradation is a result of the MPEG compression algorithm being able to only partly reconstruct the picture. Weak reception can be caused by several factors.

You may be pointing at the wrong satellite, or at a satellite whose footprint does not properly cover your geographic region. Essentially all receivers or receiving programs have a menu that shows the current satellite or the current satellite's network ID number. For example the Bell ExpressVu satellite at latitude 91°W has network ID 0256, and Sky UK has network ID 002. The table giving the network ID for many providers can be found in the relevant Appendix. If the satellite does not have a strong enough signal (which could be determined from the footprint or EIRP map), then a larger dish might be required.

Another common cause of weak reception is an obstruction between the dish and the satellite. Remember that the signal bounces off the dish to the LNB, so the direction the dish is "looking" is not directly in front of it, but tilted upwards in most cases. Trees along the line of sight will obstruct its reflection.

In some cases a dish needs to be skewed i.e. rotated sideways by some amount. Incorrect skew can cause bad reception. The skew is related to the polarization of the signal that is received, and an incorrect skew can lead to a "mixup" between the horizontal and vertical polarization. This can look like weak reception even though a simple signal strength measurement will seem to show that there is no problem.

You should also assure that the antenna cable run is not too long and that the cable is the correct type (typically RG6). If the cable is longer than 20 m (60 feet) the cable quality and placement should examined carefully. Runs of 100 m (300 feet) are common and even cable lengths of 300 m (900 feet) are sometimes possible with the right equipment, but signal quality degrades with length. If the cable is of the right type, check the quality of the connectors at the ends and assure the inner core and outer shield are not short-circuited together, that there is no rust, and that the connection from the F-connector ends to the internal wires looks secure.

If a DiSEqC switch (or related switches such as an SW21 or Digibox) is used between the LNB and receiver, it should be verified that it is working correctly. DisEqC switches, in particular, can be damaged by the weather or by connection and disconnection of the equipment while the receiver is turned on. Finally, the LNB itself should be checked since these units are also exposed to extreme weather conditions which can result in failure.

Appendix C.   Standard video formats

This is a table of some of the most common video signaling standards. Note that the number of "active pixels" in an image refers to the number that can be expected to have picture information attached to them. In some cases additional pixels may be transmitted and it is very commonplace for displays to crop the picture so that not all the active pixels are actually shown.

Name	Samples/ Line or columns or width	Active Lines or rows or height	Comments
QCIF, H.261 + H.263	176	144	
MPEG 1 SIF, PAL	352	288	Source Input Format, used for PAL VCDs
MPEG 1 SIF, NTSC	352	240	Source Input Format, used for NTSC VCDs
SIF, QVGA	320	240	
SIF, alternate, digital	384	288	1/4 of the 768x576 standard
NTSC, analog	640	480	Standard definition (SD) 525 lines total of which only 480 are visible on-screen
NTSC, analog DV, DVB, DVD, SVCD	720	480	Standard definition (SD)
D1 standard NTSC, analog CCIR	720	486	Standard definition (SD)
PAL-60 PAL-M		525	
PAL, analog PAL B PAL G/D/K/I		576	576 active lines 625 lines total (includes caption data) 25 frames per second

Name	Samples/ Line or columns or width	Active Lines or rows or height	Comments
SECAM, analog		576	576 active lines 625 lines total (includes caption data)
Enhanced definition (EDTV), digital			Same as NTSC/PAL but with progressive scan
HDTV	1280	720	Interlaced or progressive, typically interlaced for broadcast TV.
HDTV XGA	1024	768	
HDTV "True HDTV"	1920	1080	Progressive (usually)
Ultra-High, LCD TV	2160		Progressive
SMPTE 170M			NTSC composite analog
SMPTE 240M	1920	1035	Interlaced
SMPTE 244M		525	Originally designed for digitizing composite signals
SMPTE 260M	1920	1035	Interlaced
SMPTE 274M	1920	1080	Interlaced or progressive
SMPTE 296M	1280	720	Progressive
SMPTE 372M			Dual-link SDI (4:4:4)
ITU-R BT.601-4	720	483	NTSC, interlaced 480: analog, digital 4:3, digital 16:9; progressive
ITU-R BT.601-4	720	576	PAL, interlaced 480: analog, digital 4:3, digital 16:9; progressive

Appendix D.   More information

A web page containing errata and supplementary information for buyers of this book is available at

http://www.y1d.com/DTVbook/links

Some links may require the access key: **DTVAH8002**

Appendix E. Descriptor values

Descriptors are the numbers used to identify data stream packets with respect to the kind(s) of data they contain. The most important are the audio and video streams, but many others exist and are listed below. Note that some descriptor values are common to all MPEG-2 streams, others are specific to the ATSC or DVB standards, and some are proprietary. A few examples of proprietary descriptors are given here, but they are not publicly documented by the broadcasters, and are subject to change.

MPEG-2, Non-DVB	
0x02	Video Stream
0x03	Audio Stream
0x04	Hierarchy
0x05	Registration
0x06	Data Stream Alignment
0x07	Target Background Grid
0x08	Video Window
0x09	Conditional Access
0x0a	ISO 639 Language
0x0b	System Clock
0x0c	Multiplex Buffer Utilization
0x0d	Copyright Descriptor
0x0e	Maximum Bitrate
0x0f	Private Data Indicator
0x10	Smoothing Buffer
0x11	STD
0x12	IBP

DVB	
0x40	Network_information_section
0x41	Service List
0x42	Stuffing
0x43	Satellite Delivery
0x44	Cable Delivery
0x45	VBI Data
0x46	VBI Teletext
0x47	Bouquet Name
0x48	Service

0x49	Country Availability
0x4a	Linkage
0x4b	NVOD Reference
0x4c	Time Shifted Service
0x4d	Short Event
0x4e	Extended Event
0x4f	Time Shifted Event
0x50	Component
0x51	Mosaic
0x52	Stream Identifier
0x53	Conditional Access
0x54	Content
0x55	Parental Rating
0x56	Teletext
0x57	Telephone
0x58	Local Time Offset
0x59	Subtitling
0x5a	Terrestrial Delivery
0x5b	Multi Lingual Network Name
0x5c	Multi Lingual Bouquet Name
0x5d	Multi Lingual Service Name
0x5e	Multi Lingual Component Name
0x5f	Private Data Specified
0x60	Service Move
0x61	Short Smoothing Buffer
0x62	Frequency List
0x63	Partial Transport Stream
0x64	Data Broadcast
0x65	CA System
0x66	Data Broadcast ID
0x67	Transport Stream
0x68	DSNG
0x69	PDC
0x6a	AC-3 Audio
0x6b	Ancillary Data
0x6c	Cell List
0x6d	Cell Frequency Link
0x6e	Announcement Support
0x73	DTS Audio
0x83	Logical Channel Number

DCII	
0x80	Stuffing
0x81	AC-3 Audio
0x82	Frame Rate
0x83	Extended Video
0x84	Component Name
0x90	Frequency Spec
0x91	Modulation Parameters
0x92	Transport Stream ID
0xc0	Banner Override
ATSC	
0x80	Stuffing
0x81	AC-3 Audio
0x86	Caption Service
0x87	Content Advisory
0xa0	Extended Channel Name
0xa1	Service Location
0xa2	Time-Shifted Service
0xa3	Component Name
Dish Network Proprietary	
0x91	Dish Network Compressed Short Event ID
0x92	Dish Network Compressed Extended Event ID

Appendix F.   CA system ID codes

Different conditional access systems are identified by numerical codes in the relevant packets. The CA code is critical for the authorization of encrypted or restricted content, and can be used to determine if one should be able to receive the programming.

ID/Range	CA System description
0x0000	Reserved
0x0001	IPDC SPP Open Security Framework Generic Roaming
0x0002	18Crypt (IPDC SPP (TS 102 474) Annex B)
0x0003	DVB Content Protection & Copy Management
0x0004 - 0x00FF	Standardized systems
0x0100 - 0x01FF	Canal Plus
0x0200 - 0x02FF	CCETT
0x0300 - 0x03FF	Kabel Deutschland
0x0400 - 0x04FF	Eurodec
0x0500 - 0x05FF	France Telecom
0x0600 - 0x06FF	Irdeto
0x0700 - 0x07FF	Jerrold/GI/Motorola
0x0800 - 0x08FF	Matra Communication
0x0900 - 0x09FF	News Datacom
0x0A00 - 0x0AFF	Nokia
0x0B00 - 0x0BFF	Norwegian Telekom
0x0C00 - 0x0CFF	NTL
0x0D00 - 0x0DFF	CrytoWorks (Irdeto)
0x0E00 - 0x0EFF	Scientific Atlanta
0x0F00 - 0x0FFF	Sony
0x1000 - 0x10FF	Tandberg Television
0x1100 - 0x11FF	Thomson
0x1200 - 0x12FF	TV/Com
0x1300 - 0x13FF	HPT - Croatian Post and Telecommunications
0x1400 - 0x14FF	HRT - Croatian Radio and Television
0x1500 - 0x15FF	IBM
0x1600 - 0x16FF	Nera
0x1700 - 0x17FF	BetaTechnik
0x1800 - 0x18FF	Kudelski SA
0x1900 - 0x19FF	Titan Information Systems
0x2000 - 0x20FF	Telefonica Servicios Audiovisuales
0x2100 - 0x21FF	STENTOR (France Telecom, CNES and DGA)

ID/Range	CA System description
0x2200 - 0x22FF	Scopus Network Technologies
0x2300 - 0x23FF	BARCO AS
0x2400 - 0x24FF	StarGuide Digital Networks
0x2500 - 0x25FF	Mentor Data System, Inc.
0x2600 - 0x26FF	European Broadcasting Union
0x2700 - 0x270F	PolyCipher (NGNA, LLC)
0x4700 - 0x47FF	General Instrument (Motorola)
0x4800 - 0x48FF	Telemann
0x4900 - 0x49FF	CrytoWorks (China) (Irdeto)
0x4A00 - 0x4A0F	Tsinghua TongFang
0x4A10 - 0x4A1F	Easycas
0x4A20 - 0x4A2F	AlphaCrypt
0x4A30 - 0x4A3F	DVN Holdings
0x4A40 - 0x4A4F	Shanghai Advanced Digital Technology Co. Ltd. (ADT)
0x4A50 - 0x4A5F	Shenzhen Kingsky Company (China) Ltd.
0x4A60 - 0x4A6F	@Sky
0x4A70 - 0x4A7F	Dreamcrypt
0x4A80 - 0x4A8F	THALESCrypt
0x4A90 - 0x4A9F	Runcom Technologies
0x4AA0 - 0x4AAF	SIDSA
0x4AB0 - 0x4ABF	Beijing Compunicate Technology Inc.
0x4AC0 - 0x4ACF	Latens Systems Ltd
0x4AD0 - 0x4AD1	XCrypt Inc.
0x4AD2 - 0x4AD3	Beijing Digital Video Technology Co., Ltd.
0x4AD4 - 0x4AD5	Widevine Technologies, Inc.
0x4AD6 - 0x4AD7	SK Telecom Co., Ltd.
0x4AD8 - 0x4AD9	Enigma Systems
0x4ADA	Wyplay SAS
0x4ADB	Jinan Taixin Electronics, Co., Ltd.
0x4ADC	LogiWays
0x4ADD	ATSC System Renewability Message (SRM)
0x4ADE	CerberCrypt
0x4ADF	Caston Co., Ltd.
0x4AE0 - 0x4AE1	Digi Raum Electronics Co. Ltd.
0x4AE2 - 0x4AE3	Microsoft Corp.
0x4AE4	Coretrust, Inc.
0x4AE5	IK SATPROF
0x4AE6	SypherMedia International
0x4AE7	Guangzhou Ewider Technology Corporation Limited

ID/Range	CA System description
0x4AE8	FG Crypt
0x4AE9	Dreamer-i Co., Ltd.
0x4AEA	Cryptoguard AB
0x4AEB	Abel DRM Systems AS
0x4AEC	FTS DVL SRL
0x5347	GkWare e.K.
0x5601	Verimatrix, Inc.

Appendix G.   DirecTV error messages

DirecTV receivers use numerical error codes that are quite cryptic. This table explains some of most common that have been reported by the user community. These codes are anecdotal, unofficial, and may not apply to a given receiver.

Error Code	Explanation
711	Subscriber account was set-up, but the system has not yet been activated.
721	Current channel is not authorized for viewing, or else the subscription has expired.
722	A "resend" is required from DirecTV to re-activate an active subscription (i.e. The receiver was probably turned off for a long period).
731	Access Card is full of PPV entries and has not reported to DirecTV (needs to dial in).
732	PPV has not been set-up, or else has been blocked due to problem with the account.
733	Due to unsuccessful attempts by the receiver to download the access card, the PPV purchase ceiling has been set to 1 cent by DirecTV
734	Receiver not usable now for PPV purchases capability due to some receiver malfunction.
745	The smart card being used has been reported lost, stolen, etc. and is cancelled.
746	Software error.
751	The smart card needs to be upgraded a newer version.
752	The smart card is not usable with this receiver hardware

Appendix H. Original network ID values

The original network ID is a number that describes the source broadcasting network that a stream comes from. It stays the same even when the program is rebroadcast by a different broadcaster. It is like the original author of a program.

ID Range	Description	Operator
0x0000	Reserved	(Reserved)
0x0001	Astra Satellite Network 19.2°E	Société Européenne des Satellites
0x0002	Astra Satellite Network 28.2°E	Société Européenne des Satellites
0x0003 - 0x0019	Astra 1 - 23	Société Européenne des Satellites
0x001A	Quiero Televisión	Quiero Televisión
0x001B	RAI	RAI
0x001C	HELLAS SAT	Hellas-Sat S.A.
0x001D	Telecom Italia Media	TELECOM ITALIA MEDIA BROADCASTING SRL
0x001F	Europe Online Networks (EON)	Europe Online Networks S.A
0x0020	ASTRA	Société Européenne des Satellites
0x0021 - 0x0026	Hispasat Network 1 - 6	Hispasat S.A .
0x0027	Hispasat 30°W	Hispasat FSS
0x0028	Hispasat 30°W	Hispasat DBS
0x0029	Hispasat 30°W	Hispasat America
0x002A	Päijät-Hämeen Puhelin Oyj	Päijät-Hämeen Puhelin Oyj
0x002B	Digita DVB-H Mobile TV network	Digita Oy
0x002E	Xantic	Xantic BU Broadband
0x002F	TVNZ Digital	TVNZ
0x0030	Canal+ Satellite Network	Canal+ SA (for Intelsat 601-325°E)
0x0031	Hispasat – VIA DIGITAL	Hispasat S.A.
0x0032 -	Hispasat Network 7 - 9	Hispasat S.A.

ID Range	Description	Operator
0x0034		
0x0035	Nethold Main Mux System	NetHold IMS
0x0036	TV Cabo	TV Cabo Portugal
0x0037	STENTOR	France Telecom, CNES and DGA
0x0038	OTE	Hellenic Telecommunications Organization S.A .
0x0039	Broadcast Australia Datacasting	Broadcast Australia Pty.
0x003A	GeoTel	GeoTelecom Satellite Services
0x003B	BBC	BBC
0x003C -	KPN	KPN Broacast Services
0x0040	Croatian Post and Telecommunications	HPT – Croatian Post and Telecommunications
0x0041	Mindport network	Mindport
0x0042	DMG	DTV haber ve Gorsel yay_ncilik
0x0043	Tividi	arena Sport Rechte und Marketing GmbH
0x0044 -	VisionTV	VisionTV LLC
0x0045	Vision TV	SES-Sirius
0x0046	1 degree W #1	Telenor
0x0047	1 degree W #2	Telenor
0x0048	STAR DIGITAL	STAR DIGITAL A.S .
0x0049	Sentech Digital Satellite	Sentech
0x004D	Eutelsat satellite system at 4°East	Skylogic Italia
0x004E	Eutelsat satellite system at 4°East	Eutelsat S.A.
0x004F	Eutelsat satellite system at 4°East	Eutelsat S.A.
0x0050	Croatian Radio and Television	HRT – Croatian Radio and Television
0x0051	Havas	Havas
0x0052	Osaka Yusen Satellite	StarGuide Digital Networks
0x0053	PT Comunicações	PT Comunicações
0x0054	Teracom Satellite	Teracom AB Satellite Services
0x0055	Sirius Satellite System European Coverage	NSAB (Teracom)
0x0058	(Thiacom 1 & 2 co-located 78.5°E)	UBC Thailand

ID Range	Description	Operator
0x005E	Sirius Satellite System Nordic Coverage	NSAB
0x005F	Sirius Satellite System FSS	NSAB
0x0060	Kabel Deutschland	Kabel Deutschland
0x0069	Optus B3 156°E	Optus Communications
0x0070	BONUM1; 36 Degrees East	NTV+
0x0073	PanAmSat 4 68.5°E	Pan American Satellite System
0x007D	Skylogic	Skylogic Italia
0x007E - 0x007F	Eutelsat Satellite System at 7°E	EUTELSAT – European Telecommunications Satellite Organization
0x0085	BetaTechnik	BetaTechnik
0x0090	National network	TDF
0x00A0	National Cable Network	News Datacom
0x00A1	DigiSTAR	STAR Television Productions Ltd (HK) (NDS)
0x00A2	Sky Entertainment Services	NetSat Serviços Ltda (Brazil), Innova S. de R. L. (Mexico) and Multicountry Partnership L. P.
0x00A3	NDS Director systems	Various (product only sold by Tandberg TV) (NDS)
0x00A4	ISkyB	STAR Television Productions Ltd (HK) (NDS)
0x00A5	Indovision	PT. Matahari Lintas Cakrawala (MLC) (NDS)
0x00A6	ART	ART (NDS)
0x00A7	Globecast	France Telecom (NDS)
0x00A8	Foxtel	Foxtel (Australia) (NDS)
0x00A9	Sky New Zealand	Sky Network Television Ltd (NDS)
0x00AA	OTE	OTE (Greece) (NDS)
0x00AB	Yes Satellite Services	DBS (Israel) (NDS)
0x00AC	(NDS satellite services)	(NDS to be allocated)
0x00AD	SkyLife	Korea Digital Broadcasting
0x00AE - 0x00AF	(NDS satellite services)	(NDS to be allocated)
0x00B0 - 0x00B3	TPS	La Télévision Par Satellite
0x00B4	Telesat 107.3°W	Telesat Canada

ID Range	Description	Operator
0x00B5	Telesat 111.1°W	Telesat Canada
0x00B6	Telstra Saturn	TelstraSaturn Limited
0x00BA	Satellite Express – 6 (80°E)	Satellite Express
0x00C0 - 0x00CD	Canal +	Canal+
0x00D0	CCTV	China Central Television (NDS)
0x00D1	Galaxy	Galaxy Satellite Broadcasting, Hong Kong (NDS)
0x00D2 - 0x00DF	(NDS satellite services)	(NDS to be allocated)
0x00EB	Eurovision Network	European Broadcasting Union
0x0100	ExpressVu	ExpressVu Inc.
0x010D	Skylogic	Skylogic Italia
0x010E - 0x010F	Eutelsat Satellite System at 10°E	European Telecommunications Satellite Organization
0x0110	Mediaset	Mediaset
0x011F	visAvision Network	European Telecommunications Satellite Organization
0x013D	Skylogic	Skylogic Italia
0x013E - 0x013F	Eutelsat Satellite System 13°E	European Telecommunications Satellite Organization
0x016D	Skylogic	Skylogic Italia
0x016E - 0x016F	Eutelsat Satellite System at 16°E	European Telecommunications Satellite Organization
0x01F4	MediaKabel B.V	
0x022D	Skylogic	Skylogic Italia
0x022E - 0x022F	Eutelsat Satellite System at 21.5°E	EUTELSAT – European Telecommunications Satellite Organization
0x026D	Skylogic	Skylogic Italia
0x026E - 0x026F	Eutelsat Satellite System at 25.5°E	EUTELSAT – European Telecommunications Satellite Organization
0x029D	Skylogic	Skylogic Italia
0x029E - 0x029F	Eutelsat Satellite System at 29°E	European Telecommunications Satellite Organization
0x02BE	Arabsat	Arabsat (Scientific Atlanta, Eutelsat)
0x033D	Skylogic at 33°E	Skylogic Italia
0x033E - 0x033F	Eutelsat Satellite System at 33°E	Eutelsat

ID Range	Description	Operator
0x036D	Skylogic	Skylogic Italia
0x036E - 0x036F	Eutelsat Satellite System at 36°E	European Telecommunications Satellite Organization
0x03E8	Telia	Telia, Sweden
0x045D	Eutelsat satellite system at 15°West	Eutelsat S.A.
0x045E	Eutelsat satellite system at 15°West	Eutelsat S.A.
0x045F	Eutelsat satellite system at 15°West	Eutelsat S.A.
0x047D	Skylogic	Skylogic Italia
0x047E - 0x047F	Eutelsat Satellite System at 12.5°W	EUTELSAT – European Telecommunications Satellite Organization
0x048D	Skylogic	Skylogic Italia
0x048E - 0x048F	Eutelsat Satellite System at 48°E	European Telecommunications Satellite Organization
0x049D	Eutelsat satellite system at 11°West	Eutelsat S.A.
0x049E	Eutelsat satellite system at 11°West	Eutelsat S.A.
0x049F	Eutelsat satellite system at 11°West	Eutelsat S.A.
0x052D	Skylogic	Skylogic Italia
0x052E - 0x052F	Eutelsat Satellite System at 8°W	EUTELSAT – European Telecommunications Satellite Organization
0x053D	Eutelsat satellite system at 53°East	Eutelsat S.A.
0x053E	Eutelsat satellite system at 53°East	Eutelsat S.A.
0x053F	Eutelsat satellite system at 53°East	Eutelsat S.A.
0x055D	Skylogic at 5°W	Skylogic Italia
0x055E - 0x055F	Eutelsat Satellite System at 5°W	Eutelsat
0x0600	UPC Satellite	UPC
0x0601	UPC Cable	UPC
0x0602	Tevel	Tevel Cable (Israel)
0x071D	Skylogic at 70.5°E	Skylogic Italia
0x071E -	Eutelsat Satellite	Eutelsat S.A.

ID Range	Description	Operator
0x071F	System at 70.5°E	
0x077D	Skylogic Satellite System at 7°W	Skylogic Italia
0x077E - 0x077F	Eutelsat Satellite System at 7°W	Eutelsat S.A.
0x0800 - 0x0801	Nilesat 101	Nilesat
0x0880	MEASAT 1, 91.5°E	MEASAT Broadcast Network Systems SDN. BHD. (Kuala Lumpur, Malaysia)
0x0882	MEASAT 2, 91.5°E	MEASAT Broadcast Network Systems SDN. BHD. (Kuala Lumpur, Malaysia)
0x0883	MEASAT 2, 148.0°E	Hsin Chi Broadcast Company Ltd .
0x088F	MEASAT 3	MEASAT Broadcast Network Systems SDN. BHD. (Kuala Lumpur, Malaysia)
0x1000	Optus B3 156°E	Optus Communications
0x1001	DISH Network	Echostar Communications
0x1002	Dish Network 61.5 W	Echostar Communications
0x1003	Dish Network 83 W	Echostar Communications
0x1004	Dish Network 119 W	Echostar Communications
0x1005	Dish Network 121 W	Echostar Communications
0x1006	Dish Network 148 W	Echostar Communications
0x1007	Dish Network 175 W	Echostar Communications
0x1008 - 0x100B	Dish Network W - Z	Echostar Communications
0x1010	ABC TV	Australian Broadcasting Corporation
0x1011	SBS	SBS Australia
0x1012	Nine Network Australia	Nine Network Australia
0x1013	Seven Network Australia	Seven Network Limited
0x1014	Network TEN Australia	Network TEN Limited
0x1015	WIN Television Australia	WIN Television Pty Ltd
0x1016	Prime Television Australia	Prime Television Limited
0x1017	Southern Cross Broadcasting Australia	Southern Cross Broadcasting (Australia) Limited
0x1018	Telecasters Australia	Telecasters Australia Limited
0x1019	NBN Australia	NBN Limited
0x101A	Imparja Television Australia	Imparja Television Australia
0x101B -	(Reserved for	(Reserved for Australian broadcasters)

ID Range	Description	Operator
0x101F	Australian broadcaster)	
0x1100	GE Americom	GE American Communications
0x1101	MiTV Networks Terrestrial	MiTV Networks Sdn Bhd Malaysia
0x1102	Dream Mobile TV	PMSI DVB-H
0x1103 -	PT MAC	PT. Mediatama Anugrah Citra
0x1700	Echostar 2A	Echostar Communications
0x1701	Echostar 2B	Echostar Communications
0x1702	Echostar 2C	Echostar Communications
0x1703	Echostar 2D	Echostar Communications
0x1704	Echostar 2E	Echostar Communications
0x1705	Echostar 2F	Echostar Communications
0x1706	Echostar 2G	Echostar Communications
0x1707	Echostar 2H	Echostar Communications
0x1708	Echostar 2I	Echostar Communications
0x1709	Echostar 2J	Echostar Communications
0x170A	Echostar 2K	Echostar Communications
0x170B	Echostar 2L	Echostar Communications
0x170C	Echostar 2M	Echostar Communications
0x170D	Echostar 2N	Echostar Communications
0x170E	Echostar 2O	Echostar Communications
0x170F	Echostar 2P	Echostar Communications
0x1710	Echostar 2Q	Echostar Communications
0x1711	Echostar 2R	Echostar Communications
0x1712	Echostar 2S	Echostar Communications
0x1713	Echostar 2T	Echostar Communications
0x2000	Thiacom 1 & 2 co-located 78.5°E	Shinawatra Satellite
0x2014	DTT - Andorra	STA (Servei de Telecomucaciones d'Andorra)
0x2024	Australian Digital Terrestrial Television	Australian Broadcasting Authority
0x2028	Austrian Digital Terrestrial Television	ORS - Austrian Broadcasting Services
0x2038	Belgian Digital Terrestrial Television	BIPT
0x209E	Taiwanese Digital Terrestrial Television	Directorate General of Telecommunications
0x20CB	Czech Republic Digital Terrestrial Television	Czech Digital Group

ID Range	Description	Operator
0x20D0	Danish Digital Terrestrial Television	National Telecom Agency Denmark
0x20E9	Estonian Digital Terrestrial Television	Estonian National Communications Board
0x20F6	Finnish Digital Terrestrial Television	Telecommunicatoins Administratoin Centre, Finland
0x20FA	French Digital Terrestrial Television	Conseil Superieur de l'AudioVisuel
0x2114	German Digital Terrestrial Television	IRT on behalf of the German DVB-T broadcasts
0x2174	Irish Digital Terrestrial Television	Irish Telecommunications Regulator
0x2178	Israeli Digital Terrestrial Television	BEZEQ (The Israel Telecommunication Corp Ltd .)
0x217C	Italian Digital Terrestrial Television	
0x2210	Netherlands Digital Terrestrial Television	Nozema
0x222A	DTT - New Zealand Digital Terrestrial Television	TVNZ on behalf of Freeview (NZ)
0x2242	Norwegian Digital Terrestrial Television	Norwegian Regulator
0x2260 -	DTT - Philippines Digital Terrestrial Television	NTA
0x22BE	Singapore Digital Terrestrial Television	Singapore Broadcasting Authority
0x22C1 -	DTT - Slovenia Digital Terrestrial Television	APEK - Slovenia
0x22D4	Spanish Digital Terrestrial Television	"Spanish Broadcasting Regulator
0x22F1	Swedish Digital Terrestrial Television	"Swedish Broadcasting Regulator "
0x22F4	Swiss Digital Terrestrial Television	OFCOM
0x233A	UK Digital Terrestrial Television	Independent Television Commission
0x3000	PanAmSat 4 68.5°E	Pan American Satellite System
0x5000	Irdeto Mux System	Irdeto Test Laboratories
0x616D	BellSouth	BellSouth Entertainment, Atlanta, GA,

ID Range	Description	Operator
	Entertainment	USA
0x6600	UPC Satellite	UPC
0x6601	UPC Cable	UPC
0xA011	Sichuan Cable TV Network	Sichuan Cable TV Network (PRC)
0xA012	China Network Systems	STAR Koos Finance Company (Taiwan)
0xA013	Versatel	Versatel (Russia)
0xA014	Chongqing Cable	Chongqing Municipality, PRC
0XA015	Guizhou Cable	Guizhou Province, PRC
0xA016	Hathway Cable	Hathway Cable and Datacom, India
0xA017	RCN	Rogers Cable Network, USA
0xA509	Welho Cable Network Helsinki	Welho
0xA600	Madritel	Madritel (Spain)
0xA602	Tevel	Tevel (Israel) (NDS)
0xA603	Globo Cabo (to be recycled)	Globo Cabo (Brazil) (NDS)
0xA604	Cablemas (to be recycled)	Cablemas (Mexico) (NDS)
0xA605	INC National Cable Network	Information Network Centre of SARFT (PRC) (NDS)
0xF000	SMALL CABLE NETWORKS	(Small cable network network operators)
0xF001	Deutsche Telekom	Deutsche Telekom AG
0xF010	Telefónica Cable	Telefónica Cable SA
0xF020	Cable and Wireless Communication	Cable and Wireless Communications
0xF100	Casema	Casema N.V .
0xF750	Telewest Communications Cable Network	Telewest Communications Plc
0xF751	OMNE Communications	OMNE Communications Ltd.
0xFBFC	MATAV	MATAV (Israel) (NDS)
0xFBFD	Telia Kabel-TV	Telia, Sweden
0xFBFE	TPS	La Télévision Par Satellite
0xFBFF	Sky Italia	Sky Italia Spa.
0xFC10	Rhône Vision Cable	Rhône Vision Cable
0xFC41	France Telecom Cable	France Telecom
0xFD00	National Cable	Lyonnaise Communications

ID Range	Description	Operator
	Network	
0xFE00	TeleDenmark Cable TV	TeleDenmark
0xFEC0 - 0xFEFF	Network Interface Modules	Common Interface
0xFF00 - 0xFFFE	Private_temporary_use	ETSI
0xA1018 - 0xA040	(NDS satellite services)	(NDS to be allocated)

Appendix I.   Network ID values

The following table provides a list of DVB network ID values which allow the source of programming to be identified.

Network ID	Description	Type	Operator
0x0001	Astra Satellite Network 19.2° E	Satellite	Société Européenne des Satellites
0x0002	Astra Satellite Network 28.2° E	Satellite	Société Européenne des Satellites
0x0003 - 0x0019	Astra 1 - 23	Satellite	Société Européenne des Satellites
0x001A	Intelsat IS-907 at 332.5E	Satellite	Intelsat
0x001B	TrendTV	Satellite	Communication Trends Ltd.
0x001C	HELLAS SAT	Satellite	Hellas-Sat S.A.
0x001D	NRK	IPTV	NRK
0x0020	ASTRA	Satellite	Société Européenne des Satellites
0x0021 - 0x0026	Hispasat Network 1 - 6	Satellite	Hispasat S.A .
0x0027 - 0x0029	Hispasat 30°W	Satellite	Hispasat FSS
0x002A	Multicanal	Satellite	Multicanal
0x002B	Telstra Saturn Satellite	Satellite	TelstraSaturn Limited
0x002C	Orbit Satellite Television and Radio Network	Satellite	Orbit Communications Company
0x002D	Alpha TV	Satellite	Alpha Digital Synthesis S.A.
0x002E	Xantic	Satellite	Xantic BU Broadband
0x002F	TVNZ Digital	Satellite	TVNZ
0x0030	Canal+ Satellite Network	Satellite	Canal+ SA (for Intelsat 601)
0x0031	Hispasat VIA DIGITAL	Satellite	Hispasat S.A.
0x0032 - 0x0034	Hispasat Network 7 - 9	Satellite	Hispasat S.A.
0x0035	TV Africa	Satellite	Telemedia (PTY) Ltd
0x0036	TV Cabo	Satellite	TV Cabo Portugal

Network ID	Description	Type	Operator
0x0037	STENTOR	Satellite	France Telecom, CNES and DGA
0x0038	OTE	Satellite	Hellenic Telecommunications Organization S.A .
0x0039	PMSI	Satellite	PMSI (Philippines)
0x003B	BBC	Satellite	BBC
0x0040	HPT Croatian Post and Tele-communications	To be defined	HPT Croatian Post and Telecommunications
0x0041	To be defined See Wim Mooij	Satellite	Mindport
0x0042	DMG	Satellite	DTV haber ve Gorsel yayncilik
0x0044 -	VisionTV	Satellite	VisionTV LLC
0x0045	Vision TV	Satellite	SES-Sirius
0x0046 - 0x0047	1 degree W	Satellite	Telenor
0x0048	STAR DIGITAL	Satellite	STAR DIGITAL A.S .
0x0049	Sentech Digital Satellite	Satellite	Sentech
0x004D	Eutelsat satellite system at 4°E	Satellite	Skylogic Italia
0x004E	Eutelsat satellite system at 4°E	Satellite	Eutelsat S.A.
0x004F	Eutelsat satellite system at 4°E	Satellite	Eutelsat S.A.
0x0050	HRT-Croatian Radio and Television	To be defined	HRT Croatian Radio and Television
0x0051	Havas	Satellite	Havas
0x0052	PT comunicações	Satellite	PT Comunicações
0x0053	PT comunicações	Satellite	PT Comunicações
0x0054	Teracom Satellite	Satellite	Teracom AB Satellite Services
0x0055	Sirius Satellite System European Coverage	Satellite	NSAB (Teracom)
0x0058	Thiacom 1 & 2 co-located 78.5°E	Satellite	UBC Thailand
0x005E	Sirius Satellite	Satellite	NSAB

Network ID	Description	Type	Operator
	System Nordic Coverage		
0x005F	Sirius Satellite System FSS	Satellite	NSAB
0x0060	Kabel Deutschland	Satellite	Kabel Deutschland
0x0069	Optus B3 156°E	Satellite	Optus Communications
0x0070	BONUM1 36°E	Satellite	NTV+
0x0071	TV Polsat	Satellite	Telewizja Polsat
0x0073	PanAmSat 4 68.5°E	Satellite	Pan American Satellite System
0x0074	GeoTel LMI	Satellite	GeoTelecom Satellite Services
0x0075	GeoTel Express	Satellite	GeoTelecom Satellite Services
0x0076	GeoTel 3	Satellite	GeoTelecom Satellite Services
0x007D	Skylogic	satellite	Skylogic Italia
0x007E - 0x007F	Eutelsat Satellite System at 7°E	Satellite	European Telecommunications Satellite Organization
0x0085		Satellite	BetaTechnik
0x0090	National network	Terrestrial	TDF
0x00A0	CyberStar	Satellite	Loral Space and Communications Ltd. (NDS)
0x00A1	DigiSTAR	Satellite	STAR Television Productions Ltd (HK) (NDS)
0x00A2	Sky Entertainment Services	Satellite	NetSat Servios Ltda (Brazil), Innova S. de R. L. (Mexico) and Multicountry Partnership L. P.
0x00A3	NDS Director Systems	Satellite	Various (NDS)
0x00A4	ISkyB	Satellite	STAR Television Productions Ltd (HK) (NDS)
0x00A5	Indovision	Satellite	PT. Matahari Lintas Cakrawala (MLC)

Network ID	Description	Type	Operator
			(NDS)
0x00A6	ART	Satellite	ART
0x00A7	Globecast	Satellite	France Telecom (NDS)
0x00A8	Foxtel	Satellite	Foxtel (Australia) (NDS)
0x00A9	Sky New Zealand	Satellite	Sky Network Television Ltd (NDS)
0x00AA	OTE	Satellite	OTE (Greece) (NDS)
0x00AB	Yes Satellite Services	Satellite	DBS (Israel) (NDS)
0x00AC	(NDS satellite services)	Satellite	(NDS to be allocated)
0x00AD	SkyLife	Satellite	Korea Digital Broadcasting (NDS)
0x00AE - 0x00AF	(NDS satellite services)	Satellite	(NDS to be allocated)
0x00B0 - 0x00B3	TPS	Satellite	La Télévision Par Satellite
0x00B4	Telesat 107.3°W	Satellite	Telesat Canada
0x00B5	Telesat 111.1°W	Satellite	Telesat Canada
0x00C0 - 0x00CD	Canal +	Satellite, Cable	Canal+
0x00D0 - 0x00DF	Pijt-Hmeen Puhelin Oyj	Satellite	Pijt-Hmeen Puhelin Oyj
0x00D0	CCTV	Satellite, Cable	China Central Television (NDS)
0x00D1	Galaxy	Satellite	Galaxy Satellite Broadcasting (HK) (NDS)
0x00D2 - 0x00DF	(NDS satellite services)	Satellite	(NDS to be allocated)
0x00EB	Eurovision Network	Satellite	European Broadcasting Union
0x0100 - 0x0103	ExpressVu 1 - 4	Satellite	ExpressVu Inc.
0x010D	Skylogic	satellite	Skylogic Italia
0x010E - 0x010F	Eutelsat Satellite System at 10°E	Satellite	European Telecommunications Satellite Organization
0x0110	Mediaset	Satellite	Mediaset
0x011F	visAvision Satellite	Satellite	European

Network ID	Description	Type	Operator
	Network		Telecommunications Satellite Organization
0x013D	Skylogic	satellite	Skylogic Italia
0x013E - 0x013F	Eutelsat Satellite System at 13°E	Satellite	European Telecommunications Satellite Organization
0x016D	Skylogic	satellite	Skylogic Italia
0x016E - 0x016F	Eutelsat Satellite System at 16°E	Satellite	European Telecommunications Satellite Organization
0x022D	Skylogic	satellite	Skylogic Italia
0x022E - 0x022F	Eutelsat Satellite System at 21.5°E	Satellite	EUTELSAT: European Telecommunications Satellite Organization
0x026D	Skylogic	satellite	Skylogic Italia
0x026E - 0x026F	Eutelsat Satellite at 25° E	Satellite	EUTELSAT: The European Telecommunications Satellite Organization
0x029D	Skylogic	satellite	Skylogic Italia
0x029E - 0x029F	Eutelsat Satellite at 29°E	Satellite	European Telecommunications Satellite Organization
0x033D	Skylogic at 33°E	Satellite	Skylogic Italia
0x033E - 0x033F	Eutelsat Satellite at 33°E	Satellite	Eutelsat
0x036D	Skylogic	satellite	Skylogic Italia
0x036E - 0x036F	Eutelsat Satellite at 36°E	Satellite	European Telecommunications Satellite Organization
0x0378	Selectv	Satellite	Selectv
0x03E8	Telia	Satellite	Telia, Sweden
0x045D	Eutelsat satellite at 15°W	Satellite	Eutelsat S.A.
0x045E	Eutelsat satellite at 15°W	Satellite	Eutelsat S.A.
0x045F	Eutelsat satellite at 15°W	Satellite	Eutelsat S.A.
0x047D	Skylogic	satellite	Skylogic Italia
0x047E - 0x047F	Eutelsat Satellite at 12.5°W	Satellite	EUTELSAT: European Telecommunications

Network ID	Description	Type	Operator
			Satellite Organization
0x048D	Skylogic	satellite	Skylogic Italia
0x048E - 0x048F	Eutelsat Satellite at 48°E	Satellite	European Telecommunications Satellite Organization
0x049D	Eutelsat satellite at 11°W	Satellite	Eutelsat S.A.
0x049E	Eutelsat satellite at 11°W	Satellite	Eutelsat S.A.
0x049F	Eutelsat satellite at 11°W	Satellite	Eutelsat S.A.
0x052D	Skylogic	satellite	Skylogic Italia
0x052E - 0x052F	Eutelsat Satellite at 8°W	Satellite	EUTELSAT – European Telecommunications Satellite Organization
0x053D	Eutelsat satellite at 53°E	Satellite	Eutelsat S.A.
0x053E	Eutelsat satellite at 53°E	Satellite	Eutelsat S.A.
0x053F	Eutelsat satellite at 53°E	Satellite	Eutelsat S.A.
0x055D	Skylogic at 5°W	Satellite	Skylogic Italia
0x055E - 0x055F	Eutelsat Satellite at 5°W	Satellite	Eutelsat
0x0601	UPC Satellite	Satellite	UPC
0x0616	BellSouth Entertainment	Satellite	BellSouth Entertainment, Atlanta, GA, USA
0x071D	Skylogic Satellite System at 70.5°E	Satellite	Skylogic Italia
0x071E - 0x071F	Eutelsat Satellite System at 70.5°E	Satellite	Eutelsat S.A.
0x077D	Skylogic Satellite System at 7°W	Satellite	Skylogic Italia
0x077E - 0x077F	Eutelsat Satellite at 7°W	Satellite	Eutelsat S.A.
0x0800	Nilesat 101	Satellite	Nilesat
0x0880	MEASAT 1, 91.5°E	Satellite	MEASAT Broadcast Network Systems SDN. BHD. (Kuala

Network ID	Description	Type	Operator
			Lumpur, Malaysia)
0x0882	MEASAT 2, 91.5°E	Satellite	MEASAT Broadcast Network Systems SDN. BHD. (Kuala Lumpur, Malaysia)
0x0883	MEASAT 2, 148.0°E	Satellite	Hsin Chi Broadcast Company Ltd .
0x088F	MEASAT 3	Satellite	MEASAT Broadcast Network Systems SDN. BHD. (Kuala Lumpur, Malaysia)
0x1000	Optus B3 156°E	Satellite	Optus Communications
0x1001	DISH Network	Satellite	Echostar Communications
0x1002	Dish Network 61.5° W	Satellite	Echostar Communications
0x1003	Dish Network 83° W	Satellite	Echostar Communications
0x1004	Dish Network 119° W	Satellite	Echostar Communications
0x1005	Dish Network 121° W	Satellite	Echostar Communications
0x1006	Dish Network 148° W	Satellite	Echostar Communications
0x1007	Dish Network 175° W	Satellite	Echostar Communications
0x1008 - 0x100B	Dish Network W - Z	Satellite	Echostar Communications
0x1100 - 0x110F	GE Americom	Satellite	GE American Communications
0x1700	Echostar 2A	Satellite	Echostar Communications
0x1701	Echostar 2B	Satellite	Echostar Communications
0x1702	Echostar 2C	Satellite	Echostar Communications
0x1703	Echostar 2D	Satellite	Echostar Communications
0x1704	Echostar 2E	Satellite	Echostar Communications

Network ID	Description	Type	Operator
0x1705	Echostar 2F	Satellite	Echostar Communications
0x1706	Echostar 2G	Satellite	Echostar Communications
0x1707	Echostar 2H	Satellite	Echostar Communications
0x1708	Echostar 2I	Satellite	Echostar Communications
0x1709	Echostar 2J	Satellite	Echostar Communications
0x170A	Echostar 2K	Satellite	Echostar Communications
0x170B	Echostar 2L	Satellite	Echostar Communications
0x170C	Echostar 2M	Satellite	Echostar Communications
0x170D	Echostar 2N	Satellite	Echostar Communications
0x170E	Echostar 2O	Satellite	Echostar Communications
0x170F	Echostar 2P	Satellite	Echostar Communications
0x1710	Echostar 2Q	Satellite	Echostar Communications
0x1711	Echostar 2R	Satellite	Echostar Communications
0x1712	Echostar 2S	Satellite	Echostar Communications
0x1713	Echostar 2T	Satellite	Echostar Communications
0x2000	Thiacom 1 & 2 co-located 78.5°E	Satelitte	Shinawatra Satellite
0x2001 - 0x2002	Osaka Yusen Terrestrial A - B	Terrestrial	StarGuide Digital Networks
0x2003 -	KPN	Terrestrial	KPN Broacast Services
0x2004	MiTV Networks Broadcast Terrestial Network - DVB-H	Terrestrial	MiTV Networks Sdn Bhd, Malaysia
0x2005 -	PT MAC	Terrestrial	PT. Mediatama Anugrah Citra

Network ID	Description	Type	Operator
0x2006	Dominanta DVB-H Service	Terrestrial	Dominanta LLC
0x3000	PanAmSat 4 68.5°E	Satellite	Pan American Satellite System
0x3001 - 0x3100	UK Digital Terrestrial Television	Terrestrial	Independant Television Comission
0x3001 - 0x3100	German Digital Terrestrial Television	Terrestrial	IRT on behalf of the German DVB-T broadcasts
0x3001 - 0x3100	Italian Digital Terrestrial Television	Terrestrial	Italian Telecommunications Ministry
0x3101 - 0x3200	Spanish Digital Terrestrial Television	Terrestrial	CMT (Spanish Regulstor
0x3101 - 0x3200	Swedish Digital Terrestrial Television	Terrestrial	Post och Telestyrelsen
0x3101 - 0x3200	US Digital Terrestrial Television	Terrestrial	BellSouth Entertainment, Atlanta, GA, USA (on behalf of US broadcasters)
0x3101 - 0x3200	Netherlands Digital Terrestrial Television	Terrestrial	Nozema
0x3101 - 0x3200	Czech Republic Digital Terrestrial Television	Terrestrial	Czech Digital Group
0x3201 - 0x3300	Australian Digital terrestrial Television	Terrestrial	Australian Broadcasting Authority
0x3201 - 0x3300	Irish Digital terrestrial Television	Terrestrial	Irish OFCOM
0x3201 - 0x3210	Singapore Digital Terrestrial Television	Terrestrial	Singapore Broadcasting Authority
0x3201 - 0x3300	Danish Digital Terrestrial Television	Terrestrial	National Telecom Agency Denmark
0x3201 -	Estonian Digital	Terrestrial	Estonian Terrestrial

Network ID	Description	Type	Operator
0x3300	Terrestrial Television		Regulator
0x3201 - 0x3300	Swiss Digital Terrestrial Television	Terrestrial	BAKOM
0x3201 - 0x3300	DTT - Andorra Digital Terrestrial Television	Terrestrial	STA (Servei de Telecomucaciones d'Andorra)
0x3201 - 0x3300	DTT - Slovenia Digital Terrestrial Television	Terrestrial	APEK - Slovenia
0x3211	MediaCorp	Terrestrial	MediaCorp Ltd.
0x3212	StarHub Cable Vision		StarHub Cable Vision Ltd.
0x3213 - 0x3300	Singapore Digital Terrestrial Television	Terrestrial	Singapore Broadcasting Authority
0x3301 - 0x3400	Israeli Digital Terrestrial Television	Terrestrial	BEZEQ (The Israel Telecommunication Corp Ltd.)
0x3301 - 0x3400	Finnish Digital Terrestrial Television	Terrestrial	Telecommunicatoins Administratoin Centre, Finland
0x3301 - 0x3400	French Digital Terrestrial Television	Terrestrial	Conseil Superieur de l'AudioVisuel
0x3301 - 0x3400	Taiwanese Digital Terrestrial Television	Terrestrial	Directorate General of Telecommunications
0x3301 - 0x3400	Austrian Digital Terrestrial Television	Terrestrial	ORS
0x3401 - 0x3500	Belgian Digital Terrestrial Television	Terrestrial	
0x3401 - 0x3500	Norwegian Digital Terrestrial Television	Terrestrial	
0x3401 - 0x3500	DTT - New Zealand Digital Terrestrial Television	Terrestrial	TVNZ on behalf of Freeview (NZ)

Network ID	Description	Type	Operator
0xA001	H3G	Terrestrial	3lettronica Industriale S.p.A.
0xA001 - 0xA400	Tele Denmark	Cable	Tele Denmark
0xA010	Foxtel Cable	Cable	Foxtel (Australia)
0xA011	Sichuan Cable TV Network	Cable	Sichuan Cable TV Network
0xA012	CNS	Cable	STAR Koos Finance Company (Taiwan)
0xA013	Versatel	Cable	Versatel
0xA014	New Vision Wave	Cable	SKFC (Taiwan)
0xA015	Prosperity	Cable	SKFC (Taiwan)
0xA016	Shin Ho Ho (SHH)	Cable	SKFC (Taiwan)
0xA017	Gaho	Cable	SKFC (Taiwan)
0xA018	Wonderful	Cable	SKFC (Taiwan)
0xA019	Everlasting	Cable	SKFC (Taiwan)
0xA01A	Telefirst	Cable	SKFC (Taiwan)
0xA01B	Suncrown	Cable	SKFC (Taiwan)
0xA01C	Twin Star	Cable	SKFC (Taiwan)
0xA01D	Shing Lian	Cable	SKFC (Taiwan)
0xA01E	Clearvision	Cable	SKFC (Taiwan)
0xA01F	DAWS	Cable	SKFC (Taiwan)
0xA020	Chongqing	Cable	Chongqing (PRC)
0xA021	Guizhou	Cable	Guizhou (PRC)
0xA022	Hathway	Cable	Hathway (India)
0xA023 - 0xA02C	RCN 1 - 10	Cable	Rogers Cable (USA)
0xA02D - 0xA040	(NDS services)	Cable	(NDS to be allocated)
0XA040	COMCOR-TV	Cable	COMCOR-TV
0xA041 - 0xA043	Euskaltel TV On Line	Cable	Euskaltel
0xA044	Primacom	Cable	Primacom A.G
0xA045	Hong Kong CABLE TV	Cable & MMDS	Hong Kong Cable Television Limited
0xA050 - 0xA070	Cable & Wireless Optus	Cable	Cable & Wireless Optus
0xA050 - 0xA070	Cable & Wireless Communications	Cable	Cable & Wireless Communications
0xA070	ewt Network	Cable	ewt gmbh

Network ID	Description	Type	Operator
0xA12B	Telstra Saturn Cable	Cable	TelstraSaturn Limited
0xA509	Welho Cable Network Helsinki	Cable	Welho
0xA510	NOB	Cable	Dutch Broadcast Facilities (NOB)
0xA510 - 0x589	Telefonica Cable	Cable	Telefonica Cable SA
0xA511	Martens Multimedia - Cable Networks	Cable	Martens Antennen- und Kabelanlagen Gesellschaft mbH
0xA600 - 0xA640	Cable Services de France	Cable	Cable Service de France
0xA600	Telstra HFC National Network	HFC	Telstra
0xA600	Madritel	Cable	Madritel (Spain) (NDS)
0xA601 - 0xA615	Rhône Vision Cable	Cable	Rhône Vision Cable
0xA602	Tevel	Cable	Tevel Cable (Israel) (NDS)
0xA603	Globo Cabo	Cable	Globo Cabo (Brazil) (NDS)
0xA604	Cablemas	Cable	Cablemas (Mexico) (NDS)
0xA605	Information Network Centre (INC)	Cable	Information Network Centre (China) - SARFT (NDS)
0xA61F	BellSouth Entertainment	Cable	BellSouth Entertainment, Atlanta, GA, USA
0xA641 - 0xA660	Dexys	MMDS / Cable	Dexys
0xA661 - 0xA663	Est Video Communication	Cable	Video Communication
0xA664 - 0xA666	Est Video Communication Haut-Rhin	Cable	Video Communication Haut-Rhin
0xA670 - 0xA68F	SUDCABLE Services	Cable	SUDCABLE Services
0xA697 -	OMNE	Cable	OMNE

Network ID	Description	Type	Operator
0xA69F	Communications		Communications Ltd.
0xA700	Madritel	Cable	Madritel Comunicaciones
0xA701	NTL Cable Network	Cable	NTL
0xA750	Telewest Communications Cable Network	Cable	Telewest Communications Plc
0xA751 - 0xA75F	TVCabo	Cable	TV Cabo
0xA800 - 0xA8FF	UPC Cable	Cable	UPC
0xF001 - 0xF01F	Kabel Deutschland	Cable	Kabel Deutschland
0xF100	Casema	Cable	Casema
0xF101	Bosch Telecom - Cable Networks	Cable	BOSCH Breitbandnetze
0xF11F	visAvision Cable Network	Cable	European Telecommunications Satellite Organization
0xFBFC	MATAV	Cable	MATAV Israel (NDS)
0xFBFD	Telia Kabel-TV	Cable Network	Telia, Sweden
0xFBFE	TPS	Cable Networks	la Télévision Par Satellite
0xFBFF	Sky Italia	Satellite	Sky Italia SA.
0xFC00 - 0xFCFF	France Telecom Cable	Cable	France Telecom
0xFD00 - 0xFDFF	National Cable Network	Cable	Lyonnaise Communications

Index